Band 22 Dezember 1986 **12**

isotopenpraxis

AKADEMIE-VERLAG BERLIN

ISSN 0021-1915 Isotopenpraxis, Berlin 22 (1986) 12, 417–456

An unsere Autoren

Die Zeitschrift veröffentlicht Beiträge zu folgenden Themen: Kontrolle und Analyse radioaktiver und stabiler Isotope, Anwendung radioaktiver und stabiler Isotope in allen Gebieten der Forschung und Technik, der Industrie, Landwirtschaft und Medizin, Herstellung radioaktiver und stabiler Isotope sowie von markierten Verbindungen, kernelektronische Geräte für die Isotopenanwendung (Dickenmesser, Dichtemesser u. ä.) Detektoren für isotopentechnische Arbeiten und den Strahlenschutz in Isotopenlaboratorien, Meßtechnik und Strahlenschutz, Dosimetrie in Isotopenlaboratorien, kerntechnische Geräte für die Isotopenanwendung (Beton, Bestrahlungsanlagen u. ä.), Strahlenchemie, Radiochemie, ökonomische und perspektivische Fragen, Fragen der Weiterbildung.

1. Aufnahme von Manuskripten

Zur Veröffentlichung angenommen werden nur druckreife, noch nicht anderweitig publizierte Beiträge in deutscher, englischer oder russischer Sprache zum Themenbereich der Zeitschrift und zwar

Originalarbeiten
Kurze Mitteilungen
Aus der Praxis
Zusammenfassende Berichte
Tagungsberichte
Informationen.

Sind der Beitrag oder Teile daraus bereits veröffentlicht oder ist eine anderweitige Veröffentlichung vorgesehen, so muß dies der Redaktion mit genauer Angabe der Zweitveröffentlichung mitgeteilt werden. Mit der Annahme des Manuskriptes behält sich der Verlag das Recht der Vervielfältigung, Verbreitung und Übersetzung innerhalb der gesetzlichen Schutzfrist vor. Auszugsweiser Nachdruck oder photomechanische u. a. Wiedergabe sind an eine schriftliche Genehmigung des Verlages gebunden. Manuskripte, Zeichnungen und Photos von veröffentlichten Arbeiten werden nur auf den besonderen und bis zum Zeitpunkt der Veröffentlichung mitgeteilten Wunsch des Autors zurückgesandt.

Bei der unverlangten Einreichung von Manuskripten wird empfohlen, eine Befürwortung eines der Herausgeber oder Redaktionsausschußmitglieder beizufügen.

Manuskripte sind einzureichen bei der Redaktion, 7010 Leipzig, PSF 370.

2. Form der Manuskripte

Die Autoren werden gebeten:

2 Manuskripte, eines davon in Original-Maschinenschrift, einseitig und mit 2zeiligem Abstand einzureichen; gut leserliche Photokopien sind zulässig; auf dem 1. Blatt (unter Weglassung der Bezeichnung)
den Titel der Arbeit
Vorname-Initialen und Namen sämtlicher Autoren
gegebenenfalls die Institution, in der sie tätig sind
eine kurze Zusammenfassung der Arbeit von höchstens 15 Zeilen
Keywords in alphabetischer Reihenfolge und
als Fußnote die Anschrift der Institution anzugeben;

wenn möglich, Zusammenfassung, Titel und Autoren *englisch* (bei deutschem oder russischem Manuskript) bzw. *deutsch* (bei englischem Manuskript) 2fach beizufügen;

Literaturhinweise (fortlaufend numeriert), Fußnoten, Tabellen, Abbildungen in 2—3facher Veröffentlichungsgröße), Abbildungsunterschriften (getrennt von den Abbildungen und Zeichenerklärungen, Quellenangaben, falls erforderlich, und gegebenenfalls mit einem Vermerk für besondere Wünsche bezüglich der Abbildungsgröße), private und Dienstanschrift der Verfasser mit akademischen Titeln und ausgeschriebenen Vornamen auf besonderen Blättern ans Ende des Manuskriptes zu stellen;

Wünsche für den Kolumnentitel (Seitenüberschrift) bekanntzugeben;

Absätze durch Einzüge von mindestens 3 Anschlägen kenntlich zu machen; die Maßeinheiten des SI-Systems („Systeme International") zu verwenden; Korrekturen im Manuskript unmittelbar an der zu korrigierenden Stelle anzubringen;

Formeln deutlich lesbar und möglichst groß einzutragen;

Farbabbildungen und nachträgliche Änderungen oder Ergänzungen in Abbildungen zu vermeiden;

hervorzuhebende Textstellen zu unterstreichen, und zwar halbfetten Satz mit einer geraden schwarzen Linie, kursiven Satz (Schrägschrift) mit einer blauen Wellenlinie (alle Eigennamen). Sperrungen gestrichelt;

besondere Buchstaben durch farbige Unterstreichung zu kennzeichnen, und zwar Formelzeichen in

Griechisch-Kursiv (α, β, ...) rot,
Kursiv (*A*, *B*, ...) blau
halbfetter Kursiv (**A, B**, ...) 2 Linien blau (Vektoren),
halbfetter Grotesk (**A, B**, ...) 2 Linien violett (Tensoren);

bei Literaturangaben auf Vollständigkeit zu achten. In nachstehender Reihenfolge müssen angegeben sein:

a) bei *Zeitschriften*: Vorname-Initialen und Namen sämtlicher Autoren, Titel der Zeitschrift (nach „Periodica Chimica" oder möglichst wenig gekürzt), Bandnummer (unterstrichen), Jahreszahl (in Klammern), Heftnummer (nur bei Zeitschriften, die jedes Heft mit S. 1 beginnen) sowie die Seitenzahl, auf der die zitierte Arbeit beginnt.

b) bei *Büchern*: Initialen der Vornamen und Name des Autors (Herausgeber werden durch den Zusatz „ed." gekennzeichnet), voll ausgeschriebener Titel des Buches in der Originalsprache, evtl. Bandnummer, voll ausgeschriebener Name des Verlages, Nummer der Auflage (außer der ersten), Erscheinungsort(e), Erscheinungsjahr (hier nicht in Klammern setzen!) und evtl. Kapitelnummer, Seitenzahl u. ä.

c) bei *Sammelbänden, Tagungsberichten*: Es soll zusätzlich der Titel des zitierten Beitrags angegeben werden. Die Seitenzahl muß unbedingt angegeben werden. Aus dem Zitat muß hervorgehen, ob es sich um einen Tagungssammelband oder um einen noch nicht veröffentlichten Vortrag handelt. Zwischen Tagungsort und Erscheinungsjahr des entsprechenden Tagungsberichts ist zu unterscheiden.

3. Korrekturen

Der Autor erhält zwei Umbruchsabzüge, von denen einer möglichst schnell korrigiert an die Redaktion zurückgereicht werden soll. Auf Wunsch des Autors kann die Korrektur auch von der Redaktion gelesen werden.

Durch nachträgliche Manuskriptänderungen verursachte Korrekturkosten gehen, sofern sie 10% der Satzkosten übersteigen, zu Lasten des Autors.

Der Autor erhält 50 Sonderdrucke kostenlos. Weitere Sonderdrucke können bis zur Rückgabe der Korrekturabzüge auf Kosten des Autors bei der Redaktion bestellt werden.

Bezugsmöglichkeiten

Bestellungen sind zu richten

— **in der DDR** an den Postzeitungsvertrieb unter Angabe der Kundennummer des Bestellenden oder an den AKADEMIE-VERLAG BERLIN, DDR-1086 Berlin, Leipziger Straße 3—4, PF-Nr. 1233;
— **im sozialistischen Ausland** an eine Buchhandlung für fremdsprachige Literatur oder an den zuständigen Postzeitungsvertrieb;
— **in der BRD und Berlin (West)** an eine Buchhandlung oder an die Auslieferungsstelle KUNST UND WISSEN, Erich Bieber OHG, Wilhelmstraße 4—6, D-7000 Stuttgart 1;
— **in den übrigen westeuropäischen Ländern** an eine Buchhandlung oder an die Auslieferungsstelle KUNST UND WISSEN, Erich Bieber GmbH, General Wille-Str. 4, CH-8002 Zürich;
— **im übrigen Ausland** an den Internationalen Buch- und Zeitschriftenhandel; den Buchexport Volkseigener Außenhandelsbetrieb der Deutschen Demokratischen Republik, DDR-7010 Leipzig, Postfach 160; oder an den AKADEMIE-VERLAG BERLIN, DDR-1086 Berlin. Leipziger Straße 3—4.

Zeitschrift „Isotopenpraxis"

Herausgeber: Zentralinstitut für Isotopen- und Strahlenforschung der Akademie der Wissenschaften der DDR.
Verlag: Akademie-Verlag Berlin, DDR-1086 Berlin, Leipziger Straße 3—4, PF-Nr. 1233; Fernruf: 2236221 oder 2236229; Telex-Nr.: 114420;
Bank: Staatsbank der DDR, Berlin, Kto.-Nr.: 6836-26-20712.
Chefredakteur: Prof. Dr. H. Hübner, Leipzig.
Anschrift der Redaktion der Zeitschrift Isotopenpraxis: Prof. Dr. H. Hübner, DDR-7010 Leipzig, Karl-Marx-Platz, Hauptpostamt, PSF 370; Fernruf: 281018.
Veröffentlicht unter der Lizenznummer 1527 des Presseamtes beim Vorsitzenden des Ministerrates der Deutschen Demokratischen Republik.
Gesamtherstellung: VEB Druckerei „Thomas Müntzer", DDR-5820 Bad Langensalza.
Erscheinungsweise: Die Zeitschrift „Isotopenpraxis" erscheint monatlich. Die 12 Hefte einer Jahrganges bilden einen Band. Bezugspreis je Band 240,— DM zuzüglich Versandspesen; Preis je Heft 20,— DM. Der gültige Jahresbezugspreis für die DDR ist der Postzeitungsliste zu entnehmen.
Bestellnummer dieses Heftes: 1079/22/12.

AN (EDV) 12831
00900

Isotopenpraxis 22 (1986) 12, S. 417 bis 419

Originalarbeiten

Eindimensionale Näherungsformel für die Intensität vorwärtsgestreuter 60-keV-Quanten in Abhängigkeit von Aschegehalt und Flächenmasse in Rohbraunkohle

J. W. Leonhardt

(Akademie der Wissenschaften der DDR, Zentralinstitut für Isotopen- und Strahlenforschung, Leipzig)[1])

Herrn Prof. Dr. rer. nat. habil. *Klaus Fuchs* zum 75. Geburtstag gewidmet

Eine eindimensionale Näherungsformel für vorwärtsgestreute Quanten in Abhängigkeit vom Aschegehalt c_A und der Flächenmasse ϱx in Rohbraunkohle wird angegeben. Für einen realen Datensatz für Kohlen verschiedener Aschegehalte und konstanter Schütthöhen wird gezeigt, daß ϱx- oder Dichteschwankungen korrigiert werden können, wenn sie mittels 660-keV-transmittierender Quanten simultan erfaßt werden können. Die Standardabweichung des Korrelationsfehlers zwischen Vorwärtsstreusignal und Aschegehalt kann um den Faktor 2 verringert werden.

An onedimensional approximated formula for forward scattered 60 keV photons in dependence on the ash content and mass per square area in lignite is given. By means of realistic data evaluated from coals of various ash contents but constant level it could be shown, that ϱx or ϱ-changes could be corrected by means of 660-keV-Photons. The formula allows to reduce the mean standard deviation of the correlation between intensity of scattered photons and ash content by factor 2.

Keywords

ashes; attenuation; coal; correlations; density; gamma radiation; lignite; mass; numerical data; photons; scattering

1. Einführung

Die berührungslose Erfassung von Aschegehalten in Rohbraunkohle ist ein modernes meßtechnisches Problem, dessen Lösung wichtige Voraussetzung für die Prozeßregelung in der Kohleförderung und -verarbeitung ist. Häufig ist es notwendig, die Summe aller nichtbrennbaren Kohlebestandteile bei stark schwankenden Schichtdicken- und Dichteschwankungen im Meßgut zu bestimmen. Dies macht die genaue Kenntnis der Funktion Meßsignal-Summe aller Variabler notwendig. Während für transmittierende Quanten mit gutem Erfolg das exponentielle Schwächungsgesetz [1, 2], wie z. B. in KRAS 2, angewandt werden kann, ist der Zusammenhang zwischen Streuintensität und Variablen (ϱ, x, c_A) verwickelt. Das erschwert die ingenieurtechnische Umsetzung von Meßprinzipien, die auf der wichtige Vorteile bietenden Vorwärtsstreuung beruhen. Deshalb soll eine Näherungsformel hergeleitet und mittels eines realen Datensatzes in Hinsicht auf ihren Geltungsbereich getestet werden.

2. Eine eindimensionale Näherungsformel für vorwärtsgestreute Quanten

Die für die Herleitung der Formel wichtige Situation ist in Abb. 1 schematisch dargestellt. Ein paralleles Strahlenbündel der Intensität I_0 trifft auf das Meßgut der Dichte ϱ und Dicke x. Nach der Strecke y beträgt die Zahl der in

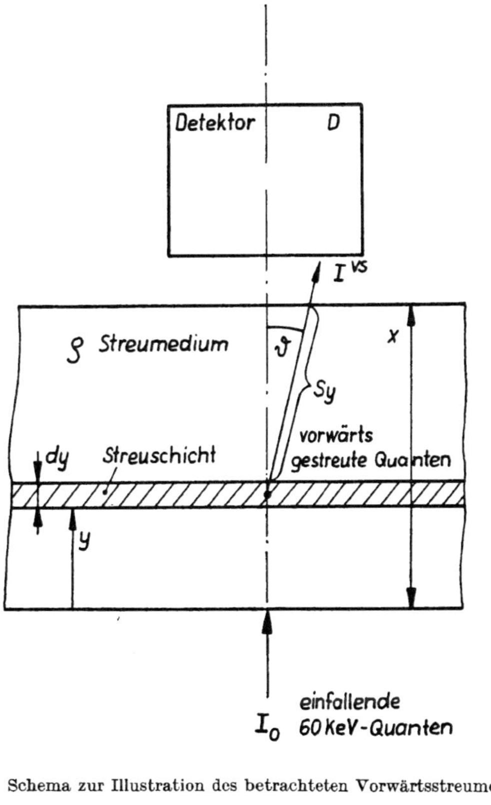

Abb. 1. Schema zur Illustration des betrachteten Vorwärtsstreumodells

[1]) Anschrift: DDR-7050 Leipzig, Permoserstr. 15

dy gestreuten Quanten

$$I_0 \mu_s^{60} \cdot \varrho \cdot e^{-\mu^{60} \varrho x} \, dy \qquad (1)$$

Darin ist μ_s^{60} der Streuanteil des Massenschwächungskoeffizienten (MSK), μ^{60} ist der totale. Für den Fall, daß für das Streumedium $\varrho x > 10$ erfüllt ist, gelangen vorzugsweise nur die unter kleinem Winkel ϑ gestreuten Quanten an den in Einfallrichtung hinter dem Streumedium positionierten Detektor D. Es wird deshalb angenommen, daß

$$\cos \vartheta = \frac{x - y}{s_y} \approx 1$$

gilt. Die Zahl der das Streumedium verlassenden gestreuten Quanten ergibt sich aus dem Integral über (1), multipliziert mit $e^{-\mu^{60} \varrho (x-y)}$, der Wahrscheinlichkeit, daß der in y gestreute Quant nicht einer weiteren Wechselwirkung unterliegt. Dieses liefert für den eindimensionalen Fall eine Näherungsformel (2), die mit relativ guter Genauigkeit die Intensität gestreuter Quanten, die einen Detektor erreichen, in Abhängigkeit von μ^{60}, ϱ, x beschreibt.

$$I^{VS} = I_0^{60} g(x) \, \mu_s^{60} \cdot \varrho \cdot x \exp(-\mu^{60} \cdot \varrho x) \qquad (2)$$

Der Geometriefaktor $g(x)$, ist systemtypisch und wird hier als Konstante behandelt, da für den Text $x =$ const. gilt. (2) berücksichtigt die Winkelabhängigkeit $\mu^{60}(\vartheta)$ nicht, ebenso nicht die durch Energieverluste bei Comptonstreuung auftretenden Änderungen in μ^{60}. Diese Annahmen sind rechtfertigbar. Nach *Clayton* [4] kann für die Aschegehaltsbestimmung der MSK eines 2-Komponentengemisches näherungsweise geschrieben werden:

$$\mu^{60} = \mu_{OM}^{60} + \mu_{AM}^{60} \qquad \begin{array}{l} OM = \text{organic matter} \\ AM = \text{anorganic matter} \end{array}$$

Für Braunkohle muß der hohe Wasseranteil c berücksichtigt werden, so daß gilt:

$$\mu^{60} = \mu_0^{60} + c_A \Delta\mu^{60} \qquad (3)$$

darin sind

$$\mu_0^{60} = c \mu_{H_2O}^{60} + (1 - c) \, \mu_c^{60}$$

$$\Delta\mu^{60} = (1 - c) \, (\mu_{eff}^{60} - \mu_c^{60})$$

Der effektive MSK, μ_{eff}^{60}, ist für eine konstante Matrix der anorganischen Bestandteile in der Kohle definiert, er ergibt sich nach [5] aus (3):

$$\mu_{eff}^{60} = \sum_i f_i \cdot \mu_i^{60}$$

Darin sind μ_i^{60} der MSK des Aschebestandteiles i, f_i ist dessen Konzentration. Unter der Annahme, die betrachtete Kohle bestehe aus 52% H_2O, die Asche aus 50% SiO_2, 40% Al_2O_3 und 10% Fe_2O_3, können die in (4) und (3) angeführten Größen berechnet werden. Sie sind in Tab. 1 zusammengefaßt. (3) wird in (2) eingesetzt. Die aktuelle Dichte wird als $\varrho = \varrho_0 \pm \Delta\varrho$ dargestellt, worin ϱ_0 die mittlere Dichte, $\Delta\varrho$ die Dichteänderung sind.

$$I^{VS} = I_0^{60} \cdot g(x) \cdot \mu_s^{60} (\varrho_0 \pm \Delta\varrho) \, x \exp$$
$$\times [-(\mu_0^{60} + c_A \, \Delta\mu^{60}) \, (\varrho_0 \pm \Delta\varrho) \, x] \qquad (5)$$

Tab. 1. Nach *Siegbahn* [5] berechnete Massenschwächungskoeffizienten

$\mu_{H_2O}^{60}$	μ_c^{60}	μ_0^{60}	μ_{eff}^{60}	$\Delta\mu^{60}$	μ^{660}	
0,137	0,174	0,155	0,300	0,126	0,084	$cm^2 g^{-1}$

Die Exponentialfunktion wird modifiziert; für Exponenten < 1 kann eine Reihe entwickelt werden:

$$e^{-\mu_0^{60} \Delta\varrho x} \approx 1 - \mu_0^{60} \Delta\varrho x$$

$$e^{-c_A \Delta\mu^{60}} (\varrho_0 \pm \Delta\varrho) \, x \approx 1 - c_A \cdot \Delta\mu^{60} (\varrho_0 \pm \Delta\varrho) \, x$$

Es läßt sich zeigen, daß $\mu_0^{60} \Delta\varrho x < 1$ und $c_A \Delta\mu^{60} \varrho_0 x < 1$ für die meisten interessierenden Fälle erfüllt sind. Somit ergibt sich:

$$I^{VS} = I_0^{60} g(x) \, \varrho_0 x \mu_s^{60} \, e^{-\mu_0^{60} \varrho_0 x} \, (1 - c_A \Delta\mu^{60} \varrho x)$$
$$\times \left\{ 1 + \mu_0^{60} (\varrho x - \varrho_0 x) \left[\frac{1}{\varrho_0 x \mu_0^{60}} - \frac{\varrho x - \varrho_0 x}{\varrho_0 x} \right] \right\} \qquad (6)$$

Formel (6) gibt den Algorithmus für die Korrektur in bezug auf Dichteänderungen, $\Delta\varrho$, in I^{VS}. Es läßt sich zeigen, daß (6) weiter vereinfacht werden kann, wenn man beachtet, daß mit relativ guter Genauigkeit

$$\left\{ 1 + \mu_0^{60} (\varrho x - \varrho_0 x) \left[\frac{1}{\varrho_0 x \mu_0^{60}} - \frac{\varrho x - \varrho_0 x}{\varrho_0 x} \right] \right\} \approx 1$$

erfüllt sein muß. Es ergibt sich schließlich ein gut praktizierbarer Ausdruck:

$$I^{VS} = I_0^{60} g(x) \, \mu_s^{60} \varrho_0 x \, e^{-\mu_0^{60} \varrho_0 x} \, (1 - c_A \Delta\mu^{60} \varrho x)$$

ϱx kann mittels transmittierender 660-keV-Quanten gut dargestellt werden:

$$\varrho x = \frac{1}{\mu^{660}} \ln \left(\frac{I_0^{660}}{I^{660}} \right). \qquad (8)$$

Wenn (8) in (7) eingesetzt, gilt endlich

$$I^{VS} = I_0^{60} g(x) \, \mu_s^{60} \varrho_0 x \left[1 - c_A \, \Delta\mu^{60} \cdot \frac{1}{\mu^{660}} \right.$$
$$\left. \times \ln \left(\frac{I_0^{660}}{I^{660}} \right) \right] e^{-\mu_0^{60} \varrho_0 x} \qquad (9)$$

3. Test der Formel (9)

In einer speziellen Meßvorrichtung wurde ein Rohbraunkohlestrom konstanter Schichtdicke, $x = 20$ cm, und verschiedenen Aschegehalten, c_A, erzeugt. Die radiometrischen Signale, d. h. die Intensitäten für transmittierende 660-keV-Quanten sowie für 60-keV-vorwärtsgestreute, wurden simultan gemessen. Abb. 2 zeigt die mittels der 660-keV-Quanten gewonnenen ϱx-Werte, korreliert mit Ascheanalysewerten. Abb. 3 enthält die Intensitäten der 60-keV-vorwärtsgestreuten Quanten, korreliert mit den c_A-Werten (Kreise). Werden die Daten aus Abb. 2 entsprechend in die hergeleitete Formel (9) eingesetzt, so ergeben sich die in Abb. 3 korrigierten Meßwerte.

Der Vergleich der linearen Regression des Datensatzes mit und ohne Korrektur nach Formel (9) enthält Tab. 2.

Tab. 2. Daten der linearen Regression für die korrigierte und unkorrigierte 60-keV-Vorwärtsstreuung

Lineare Regression I^{VS} (unkorrigiert)		I^{VS} (mit (8) korrigiert)	
$I^{VS}(0)$:	313,4	$I^{VS}(0)$:	322
m:	−2,218	m:	−2,628
r:	0,985	r:	0,987
σ:	±3,6 %	σ:	±1,3 %

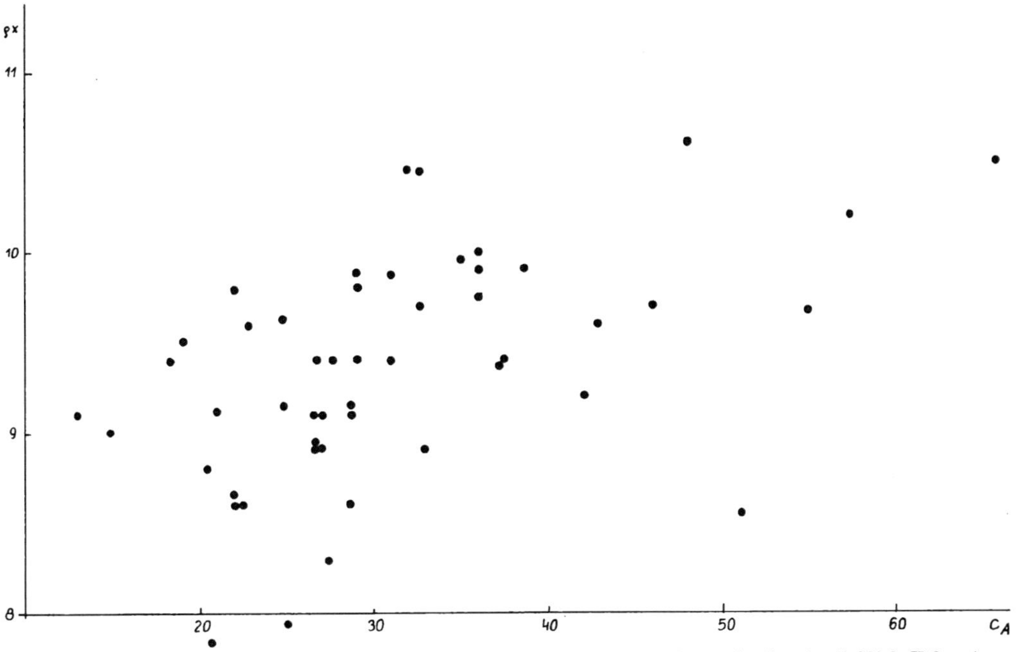

Abb. 2. Experimentell ermittelte ϱx-Werte für Rohbraunkohle verschiedener Aschegehalte c_A, bestimmt mit 660-keV-Quanten

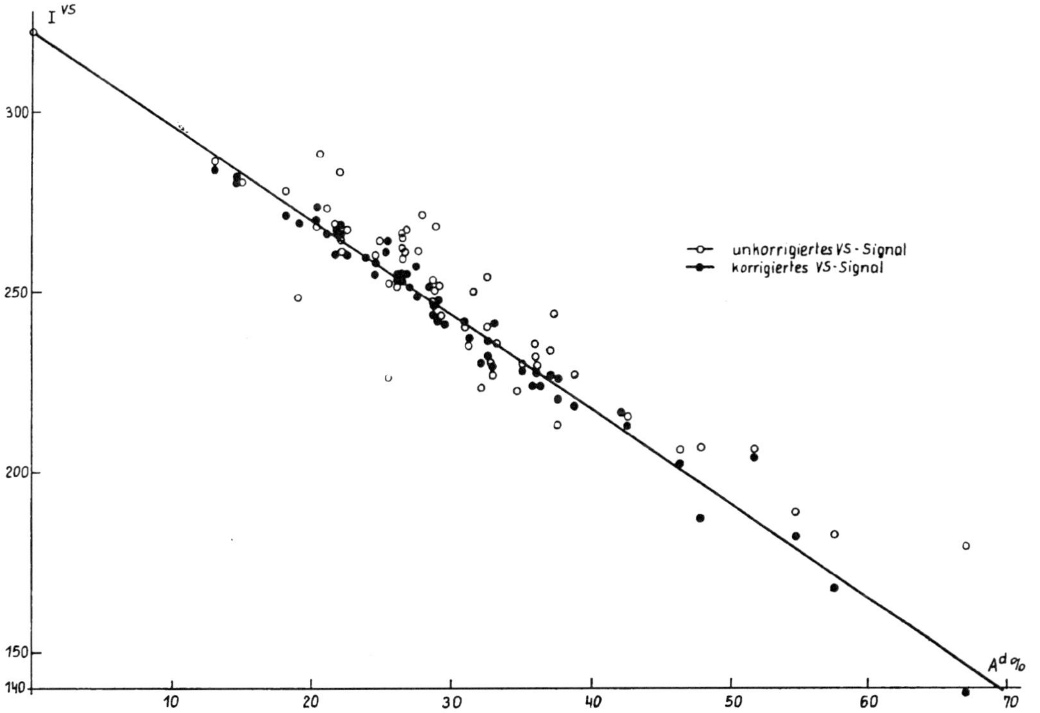

Abb. 3. Intensitäten der 60-keV-vorwärtsgestreuten Quanten für Rohbraunkohlen verschiedener Aschegehalte c_A

4. Diskussion des Ergebnisses

Anhand des statistischen Testes mit 50 Wertepaaren eines unter realen Bedingungen gewonnenen Datensatzes $I^{VS} - c_A$ konnte die Eignung der hergeleiteten Näherungsformel getestet werden. Die Korrektur von ϱx-Schwankungen im Meßgut führte zu einer Verringerung der Standardabweichung des Korrelationsfehlers der Vorwärtsstreuung von $\sigma = 3{,}6\%$ auf $\sigma = 1{,}3\%$, d. h., es ergibt sich eine Verbesserung von 2,5.

Eingegangen am 4. April 1986

Literatur

[1] *G. Körner* et al., Proceedings of 3rd Working Meeting on "Radioisotope Application and Radiation Processing in Industry" Leipzig, 1985, ZfI, Leipzig 1986, p. 819

[2] *H.-W. Thümmel, G. Körner, W. Riedel, U. Gräfenhain, Ch. Beutel*, ZfI-Mitt. **112** (1986) 23

[3] *J. W. Leonhardt, D. Fritsche, G. Körner, H.-W. Thümmel*, Isotopenpraxis **21** (1985) 352

[4] *C. G. Clayton*, in: Industrial Application of Radioisotopes and Radiation Technology, Proc. Series – STI/PUB/598, IAEA, Vienna 1981, p. 11; *I. S. Boyce, C. G. Clayton, D. Page*, in: Nuclear Techniques and Mineral Resources 1977, Proc. Series – STI/PUB/464, IAEA, Vienna 1977, p. 135

[5] *K. Siegbahn*, Beta- and Gamma Ray Spectroscopy, Amsterdam: North Holland Publ. Comp. 1955

Isotopenpraxis 22 (1986) 12, S. 420 bis 424

Precise Radiometry: Some Recent Aspects of Fruitful Interaction with Atomic Physics

G. Brunner

(Akademie der Wissenschaften der DDR, Zentralinstitut für Isotopen- und Strahlenforschung, Leipzig)[1])

Dedicated to Prof. Dr. rer. nat. habil. *Klaus Fuchs* on the occasion of his 75th birthday

Modern radiometric analytics demands a complex consideration of nuclear and electron shell processes, if more pretentious aims are envisaged. As an example the small variation of decay rates of radionuclides presents possibilities for information on chemical situations of decaying atoms. In principle this phenomenon is well known since many years, but now the situation is such that, e.g. in ^{99m}*Tc internal conversion, a full agreement of the difficult experiments and the respective theory was established. The secondary emission of X-rays as a consequence of high excitation of electron shells in combination with nuclear transitions supplies another example for a methodical progress of radiometry. Investigations on* ^{51}*Cr as an electron capture nuclide have shown that chemically induced variations of the* K_α *to* K_β *X-ray intensity ratio is at least qualitatively understood.*

Die moderne radiometrische Analytik verlangt eine komplexe Betrachtung von Kern- und Elektronenhüllen-Prozessen, sofern man anspruchsvollere Ziele anstrebt. Beispielsweise enthalten die kleinen Variationen der Zerfallsraten von Radionukliden Informationen über die chemischen Zustände der zerfallenden Atome. Im Prinzip ist diese Erscheinung seit vielen Jahren wohlbekannt, aber jetzt ist die Situation so, daß z. B. bei der inneren Konversion von ^{99m}*Tc, volle Übereinstimmung der schwierigen Experimente mit der entsprechenden Theorie erreicht wurde. Die sekundäre Röntgenemission als Folge hochangeregter Elektronenhüllen in Verbindung mit Kernübergängen liefert ein weiteres Beispiel für einen methodischen Fortschritt in der Radiometrie. Untersuchungen über* ^{51}*Cr als Nuklid mit Elektroneneinfang haben gezeigt, daß die chemisch herbeigeführten Variationen des* K_α *zu* K_β *Röntgen-Intensitätsverhältnisses zumindest qualitativ verstanden werden.*

Keywords

age estimation; atomic models; decay; electron capture; half-life; internal conversion; nuclear physics; polarization; structural chemical analysis; X radiation

1. Introduction

Radiation emission by radionuclides is an interesting phenomenon from the point of view of atomic properties. The physics of the structure of matter has shown that, no doubt, isolated studies of qualities of atomic nuclei and atomic electron shells as constituents of atomic systems represent a simple possibility which is often well adapted to the purpose of investigation.

As far as the physico-chemical effects within the outer and inner regions of the atomic electron shells are concerned it is even more important not to overlook a certain interaction, but the treatment of bound systems is very well able to distinguish between the more or less indifferent core and the valence states which are subject to a much stronger variation.

In principle, one could carry on this idea as desired: The investigation of material structures demands the flexible integration of any partial system into the hierarchy of superordinate and subordinate systems if, for methodical reasons, the definition of such a partial system is justified from the physical point of view, and it is in any case a question of the approximation desired and necessary in the given connection, how the "break-up" is carried out.

In the subsequent considerations such investigations are communicated which for reasons of the experimental methods available refer more to transition rates than to transition energies, which accounts for the circumstances of "classical" radiometry in contrast to real spectroscopy. At first questions of nucleus/electron shell interaction will stand in the foreground. Secondly, such experiments and reflections are added which refer more to the internal connections of inner and outer electron shells in the light of the electron capture process (EC) of the nucleus.

2. On Modifications of Transition Rates by Nucleus/Electron Shell Coupling in Radioactive Decay

Radioactivity is the term used to describe those spontaneous energy-releasing atomic transitions that involve changes in state of the nuclei of atoms [1]. This more recent definition has performed the transition from the formerly accepted exclusive reference to the nuclear transitions towards the characterization of the atomic situation which is including the shell, and this should be understood not only in the restricted sense of magnetic interaction of electrons and nuclear spins, which is the essential feature in hyperfine interactions, but much more emphasized with respect to the universal and consistent treatment of tran-

[1]) Address: DDR-7050 Leipzig, Permoserstr. 15

sitions particularly considered from an energetic point of view of the system nucleus/shell electrons/electromagnetic field or emitted leptons.

It is evident that labelling a nuclide as "radioactive" is influenced by these considerations to such an extent, to which the balance of the tremendous energy contents of those nuclei which contribute to a considered transition as initial and final nuclei lead to the same order of magnitude as possible excitation energies of the respective electron shells.

In this connection reference is sometimes made to the transition $^{187}Re \xrightarrow{\beta^-} {}^{187}Os$, which has even obtained a certain practical importance in extremely long range geochronology. In this β^- decay the transition energy amounts to 2·6 keV. But because the daughter nuclide exhibits a sum of all (free atomic) electron binding energies larger than that of the parent nuclide by 15·3 keV, the nuclear mass of just this parent nuclide must be 15·3 − 2·6 = 12·7 keV lower than that of the daughter nuclide in equivalent energy. Thus the energy output Q [keV] $= E_{B,Os} - E_{B,Re} - 12 \cdot 7$ of this β^- process becomes zero or even negative, if the electron binding energy in the Os atom, $E_{B,Os}$, is lowered by at least 12·7 keV. Consequently, the decay would be entirely impossible if the originating Os atom, in contrast to the initial atom was ionized in its K-shell ($E_K = 73 \cdot 860$ keV) or in its L-shell ($E_{LI} = 12 \cdot 965$ keV), respectively. But, with a certain probability such an ionization is immediately connected with the β transition itself. This is due to the modification by the decay of each electron state in the atom, ψ_{nl}. Speaking in terms of potential models: since each electron experiences different central charges by one unit ($Z' - Z = \Delta Z = 1$ in β^- emission) prior to and after decay and thus different potentials, the states $\psi_{nl}(Z)$ and $\psi_{nl}(Z')$ will have "projections" one upon another which differ from unity by finite although small amounts. Therefore

$$P_{nl} = 1 - |\langle \psi_{nl}(Z')| \psi_{nl}(Z)\rangle|^2 \tag{1}$$

yields the probability for removal of a single electron in the state ψ_{nl} during decay, that means for "shaking" it off (or up). For inner-shell orbitals P_{nl} can easily be calculated in closed form and with sufficient accuracy by hydrogenlike wave functions [2]:

$$\left.\begin{aligned} P_{2\times 1s} &\approx 2 \times \frac{3}{8}\left(\frac{\Delta Z}{Z}\right)^2 \\ P_{2\times 2s} &\approx 2 \times \frac{9}{8}\left(\frac{\Delta Z}{Z}\right)^2 . \end{aligned}\right\} \tag{2}$$

As can be seen, this decay is inhibited in net result by a shell mechanism, as the transitions "changing the nuclear charge plus K-shake" as well as "changing the nuclear charge plus L-shake" are energetically excluded. Here, the shake process presents itself as a transition without any particular operator of interaction, therefore it must be interpreted as a "monopol transition" in the simple spherical potential.

In Tab. 1 some mainly hypothetical transitions are listed which should be characterized by only very small energetic effects of the order of electron binding energies, examined by a table of atomic weights [3], being as precise as possible. The slightly negative decay energies mean: after all, such transitions are only made possible by shake processes; the low positive energies mean, as in the case of Re/Os, inhibition of decay by shake.

Tab. 1. A few selected low-energy transitions: decay into strongly ionized state is

possible		impossible	
transition	energy [keV]	transition	energy [keV]
$^{163}Dy \xrightarrow{\beta^-} {}^{163}Ho$	−9 ± 1·5	$^{187}Re \xrightarrow{\beta^-} {}^{187}Os$	2·60 ± 0·10
$^{193}Ir \xrightarrow{\beta^-} {}^{193}Pt$	−61·2 ± 3·0		
$^{205}Tl \xrightarrow{\beta^-} {}^{205}Pb$	−44 ± 4		
$^{123}Sb \xrightarrow{\beta^-} {}^{123}Te$	−57·1 ± 2·4		

Of more importance than the numerical estimate seems to be the insight that the borderline of "radioactive" and "stable" becomes diffuse to a certain extent by a comprehensive and unique consideration of the total system of nucleus and electron cloud. In doing so, by the way, one should give up a point of view of consecutive treatment as, for instance, the following: since a certain transition seems to be possible only by an electron-shell excitation of the daugther nuclide, it will not happen as far as this excitation occurs as a consequence of the nuclear transition. Rather as in unique scattering theory, the "one step" examination of the projections of the total state, containing the nucleus as well as the electron cloud, prior to and after a process under consideration, should furnish the decisive criterion whether or not the total process takes place and with what probability.

In this connection reference should be made to the significance of transition energies of a few keV for heavy nuclei in view of, e.g., the "electromagnetic" coulomb term $E_c \approx 0 \cdot 6\, Z^2\, A^{1/3}$ of the liquid drop model: referring to $E_c \approx 600$ MeV for such heavy nuclei the Re/Os-transition energy of 2·6 keV means a balance in the 10^{-6} level. On the whole, the basic principles in nuclear physics permit a monotonous connection of transition energy and transition rate. Accordingly, the nuclides of that "shadow region of radioactivity" are very long-living, if their decay is traceable at all.

Sometimes, in this connection, the question has been raised about historicity of coupling constants, such as the fine-structure constant α for electromagnetic interactions or other natural constants constituting it, e.g. *Planck's* constant [4]. Such discussions have been going on for half a century now [5]. There is, however, very little unambiguous material which could help overcome the stage of hypotheses, although the careful analysis of isotope abundances [6] and the interpretation of systematic age discordances [7] can at least supply certain indications. This topic which is very interesting although experimentally hardly tangible, finds many, rather topical, present-day continuations involving even consequences for biological evolution [8]. With some certainty one may proceed from the opinion that the interpretation of inhibition or acceleration of radioactive decay, supported by measurements of suitably selected nuclides, will further contribute to the solution of this problem.

Somewhat more than a dozen years ago *Crasemann* published a concise paper on possible atomic effects in nuclear transitions [9]. Our own investigations, referring to this, were introduced by a review containing all results that had been obtained by 1976 [10]. From this we extracted a suitable starting point for experimental work on direct influences by atomic electrons on transition rates in the field of higher than first-order processes of γ-transitions (without taking into consideration double-decay processes). Accounting for potential applications from the very beginning, the investigations concentrated mainly on technetium as an element of great interest from the point of view of catalysis and nuclear medicine.

A prerequisite for this is the necessary theoretical instrumentation, at least for the "undistorted" or atomic situation. Therefore, a program has been written [11] for the calculation of the relative probabilities of processes upwards of second order (this is the lowest order of internal conversion (IC)), which essentially supplies numerical solutions of the Dirac equation for all shells in a given atomic potential. In contrast to the well-known and otherwise very similar program CATAR [12] for the internal conversion coefficients (ICC), it does neither regard any dynamic effects of nuclear structure nor contributions out of the inner-nuclear region, indeed, but it just contains the corrections of higher order according to a theoretical ansatz of *Krutov* [13]. Its application for the second order process, IC, of ^{99m}Tc, when checking the proportionality ICC $\approx |\psi(0)|^2$ which is not at all evident, led at first to the perhaps surprising fact that the variation of the chemical or the ionization state imparts to the ICC not only via bound states, but also via states of the continuum [14]. This is especially the case for lower kinetic energies and inner shells, respectively. The "electron bridge" of γ-deexcitation as a third order effect, characterized by a virtual exchange of photons between nucleus and electron cloud, results in influences on the γ-rate of eventually many orders of magnitude for highly converted, low-energetic transitions, as we learned from the explicite numerical treatment of the E3-transition (73 eV) of ^{235m}U, the first case of full numerics of "electron bridge"-mechanism on the whole [15]. On account of the extremely high conversion degree in this example, the absolute effect remains, of course, low.

In order to investigate environmental influences, the available arsenal for describing the electronic structure first of all of atomic or free-ionic cases has been extended by the Xα scattering wave formalism [16], which also allows a cluster approximation in the sense of a multicentre approach by joining single spherical potentials of the cluster atoms together with averaged connecting regions within a common *Watson*-type sphere, carrying a smeared-out net charge of the cluster. Using in connection with this the multi-centre theory further developed and carefully refined by *E. Hartmann* for the investigation of chemical influences on inneratomic properties, the following picture emerges for the internal conversion of ^{99m}Tc [17]:

The transition energy of 2·17 keV is so low and the degree of conversion so high that the probability of decay is virtually determined by IC alone and that only electrons of the upper core and the valence orbitals can participate in the IC process. Already the atomic case, as a reference, points to a strong participation, in IC, of p-electrons, and the free-ionic treatment indicates an overcompensation of the direct IC defect due to outer (d-) ionization. This is mainly caused by 4p-charge as a reorganizational effect due to valence charge flowing off and thus core contraction.

Details are given in Tab. 2. For all halogeno complexes treated the ligands are arranged at the vertices of an octahedron (as well as for the cyano complex, too), whereas the technetate is of tetrahedral configuration. Very original is the description of the metallic-state Tc as a cluster formed by thirteen Tc atoms in a D_{3h} geometry [18].

The transformation of charge fractions to ICC's, as a result of the Xα-formalism, has been arranged by the atomic conversion program. All communicated IC variations refer to the free-atomic situation; p_{val} labels the relative ICC variation which is exclusively caused by p-type valence charge (means 5p in character) and is not present for the unbound atom. Speaking in terms of ligand-field theory, this phenomenon appears as polarizational effect of valence charge due to the field of ligands. Correspondingly, d_{val} is constituted by the respective bonds with their specific symmetry. In contrast to this, the label 4p indicates a contribution of reorganization and of "squeezing" in the upper core under the action of the ligands. Analogue influences on the inner core states 3p and 3d are relatively insignificant.

Tab. 2. Chemical variations of $\Delta\lambda/\lambda$ relative to the free atomic situation and to TcO_4^-, respectively

complex	$\Delta\lambda/\lambda$ theoretical [‰]					$\Delta\lambda/\lambda$ experimental [‰]
	p_{val}	d_{val}	4p	Σ		
	relative to the free atom				relative to TcO_4^-	
TcO_4^-	6·5	−0·4	−2·3	3·8	0	0
TcF_6^{2-}	5·1	−0·2	−1·7	3·2	−0·6	not measured
$Tc(CN)_6^{5-}$	3·7	−0·1	−1·8	1·8	−2·0	not measured
$TcCl_6^{2-}$	3·7	0·8	2·7	1·8	−2·0	−1·93 ± 0·12
$TcBr_6^{2-}$	3·4	1·3	−3·2	1·5	−2·3	−2·50 ± 0·41
TcI_6^{2-}	3·2	1·7	−4·0	0·9	−2·9	−3·30 ± 0·30
$Tc_{metall.}$	4·1	2·3	−6·0	0·4	−3·4	−3·74 ± 0·46

The theoretical net effect is in good agreement with our experiments, which were carried out extremely carefully. There is no contradiction between them. This can be seen as an extraordinary success, because it became possible that the complicated experiment which is very demanding as far as the preparation and the developed and applied radiation measurement technique are concerned, and the cluster treatment, having been developed and used for such problems for the first time, really fructified each other and proved to be consistent [19]. In the field of decay constant variation this situation was brought on for the first time. Certain preparative problems of the hexachloro-complex of Tc were surprisingly disclosed by a comparison between theory and experiment [20]. The interpretation by "squeezing"-effect only, given by *J. C. Slater* [21] for the first observations [22] of ^{99m}Tc decay constant variations sometimes ago could be recognized as insufficient.

Of course, the projection of the ^{99m}Tc-chemism on the decay rate scale with its one-dimensional character leads to restricted possibilities for structural information drawn from a $\delta\lambda/\lambda$ experiment: the measurement supplies a necessary but in most cases not a sufficient condition for structural description. But there are cases of simple alternatives for possible structures, as it proves right for the bis(dithiolato)oxo-Tc(V) complex, which is of great importance in nuclear medicine. The problem was settled in the so-called carrierfree region (10^{-8} M), where all traditional methodes failed and where the decision could come and had to come by a $\delta\lambda/\lambda$ measurement [23]. Therewith, it was again possible to extend the maximum fruitful interaction, achieved at first for methods of the sophisticated theory of multicentre description on the one hand and highly precise measurement techniques on the other into the region of immediate chemical and preparative practice in a qualitatively new manner. Of course, the treatment of Tc clusters, which was carried out within this framework, supplied spectroscopical terms of different kinds of bonding. Other working groups, having electron spectroscopical means at their disposal, have been able to verify our theoretical data on fractions of conversion electrons, stemming from diverse valence orbitals (see TcO_4^{2-} in [24]), or to use them, in their turn, for structural analytics with the help of conversion electron spectroscopy and with the tendency of a new methodical proposal for a carrier-free radioactive material [25].

Our measurements of the chemical decay-rate variations of iodine compounds by the electron-capture nuclide ^{125}I are even more demanding with regard to measuring techniques of radiation: as in the ^{99m}Tc case a quasi-differentional arrangement has been applied [26], in that two samples of different chemical bonding, to be compared with one another, are alternatively brought into the measuring channels of two scintillation counters. But, in contrast to the ^{99m}Tc case, criteria for the pulse-height selection of the 27·5 keV K X-ray of the daughter element Te as well as of the final 35·5 keV γ-radiation are inserted. As a novelty, the low penetration ability of these quantum radiations enables the alternating blocking of measuring channels for the compaired preparations not by their mechanical movement, but by opening or closing relatively thin shutters. Additionally, the precise measurements demand a reliable stabilization of pulse heights.

As low as the effect appears of e.g. $|\Delta\lambda/\lambda| = (1{\cdot}84 \pm 0{\cdot}36) \cdot 10^{-4}$ of $Na_2H_3{}^{126}IO_4$ in relation to $NaIO_3$, such significant is the contribution for the gauging problem of isomer shift in Moessbauer spectroscopy via changes of electron density near the iodine nucleus. These radiometric precision measurements stimulated further theoretical discussions on the role of exchange and overlap effects in our ([2], [27]) as well as in other working groups [28].

3. Variations of X-Ray Transition Rates by Inner-Shell/Outer-Shell Coupling, Illustrated by the Electron-Capture Decay of ^{51}Cr

Coupling between inner-shell region and valence or even supravalence orbitals is a long and well known phenomenon of which full use is made in analytics. It ranges from "chemical shifts" of transitions between (practically) pure core orbitals which may be described as hydrogenlike by chemically modified "effective" central charges, to EXAFS and its interferences of outgoing photoelectron waves partially scattered by ligands and other next neighbours.

Electron capture (EC), as a special mode of β-decay, differs from "classical" X-ray or electron spectroscopy methods in such applications as follows:

- it appears as an especially "gentle" method for such processes of necessary inner-shell excitation, which in turn can be studied in a relatively undistorted manner in using EC;
- there are no other means to get such low backgrounds in X-ray spectroscopy, that is why it is especially suited to investigations of most precise line ratios such as the intensity ratio $R_{\beta\alpha} = I_\beta/I_\alpha$;
- it represents the immediate connection to radiometry in the sense of radionuclide techniques, or in other words: the foreseeable applications are a real methodological extension of radionuclide techniques.

According to the *Soddy-Fajans*-transition rules, EC as an energetically favoured alternative to positron emission, indeed leads to the neighbouring element of lower atomic number, but charge conservation within the region of nucleus and inner core (charge equalization of proton and captured electron) results in potentials of outer electrons virtually unchanged by EC, so that a relaxation of valency happens very slowly in relation to X-ray transitions. For this reason nearly no shake appears in EC, and the X-ray emission of vanadium, as a result of ^{51}Cr-EC, fully represents the situation of valence orbitals of the parent atom Cr.

The "radiometric" situation is the following: seen for the energy dispersive spectrometry, which has been particularly developed for nuclear radiation, energy resolution for characteristic X-rays amounts to 200 ± 50 eV, depending on the quality of the usually applied semiconductor detector. According to the situation of dipole transitions that is sketched in Fig. 1, neither the K_α doublet nor the doublet $K_{\beta 1,3}$ can be resolved. A further multiplet splitting occurs for the latter one in (3d-) transition-element atoms due to the coupling of a partially filled (3d-) subshell and the vacancy of 3p-type. This causes the low-energy "chemical satellite" $K_{\beta'}$. Another very interesting $K_{\beta 5}$ component results from ligand-like valence-charge fractions if the emitting atom is in a bound situation. It can be interpreted as an internal molecular deformation or polarization of the atomic cloud by the ligand field, and it should be positively correlated to the covalent character of the bonding.

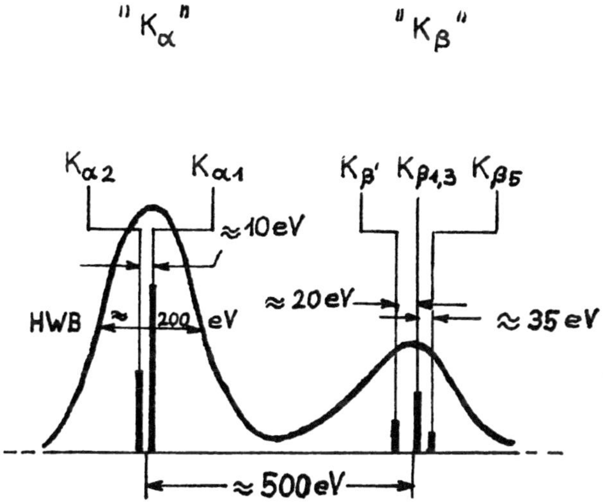

Fig. 1. K_α and K_β "lines" of an energy dispersive X-ray spectrometer and their components in the region of 3d elements

We performed precise experimental investigations of $R_{\beta\alpha}$ in addition to numerous international but surprisingly scattering results (see our summary [29]). They are based upon careful execution of measuring techniques and accompanying preparative and analytical procedures, including UV-VIS and IR spectroscopy for identification of chemical species.

Tab. 3 contains some new $R_{\beta\alpha}$-values for V X-rays following the ^{51}Cr decay. Under the impression of a successful treatment of modifications of $R_{\beta\alpha}$ in EC induced X-ray transitions of 3d-elements in terms of 3p-screening by varying 3d-charge [30], we developed a respective semi-empirical model [31]. Certain influences of polarization by ligands have been included in so far as the (de-)localized screening charge in this single-centre consideration rests upon an electronegativity [32] which depends on the oxidation number of the atom. Results of this model are given

Tab. 3. Experimental, semiempirical and theoretical K_β (all components) to K_α (all components) X-ray intensity ratios, $R_{\beta\alpha}$, for the configurations CrO_4^{2-} (Cr VI) and $CrCl_6^{3-}$ (Cr III); reference: free Cr atom

experim. (this work) $R_{\beta\alpha}$	theoretical (Xα SW)$_{rel.}$ *Band* et al.	*Hartmann*	semiemp., see [31]
CrO_4^{2-} (Cr VI)	reorganization:		
	$\Delta K_{\beta1,3}/K_{\beta1,3}$ [%]		
	1·7	1·5	
	polarization:		
	$\Delta K_{\beta5}/K_{\beta1,3}$ [%]		$\varepsilon_{Cr VI} = 2{\cdot}4$, $\varepsilon_0 = 3{\cdot}5$ } ionizity $= 0{\cdot}26$
	4·0	3·6	
	total: $\Delta K_\beta/K_{\beta1,3}$ [%]		$6 \times 0{\cdot}26 = 1{\cdot}57$ ch. u. delocalized
	5·7	5·1	
	$R_{\beta\alpha}$ (related to Scofield)		
0·1383 ± 0·0002	0·141	0·141	0·137
$CrCl_6^{3-}$ (Cr III)	reorganization:		
	$\Delta K_{\beta1,3}/K_{\beta1,3}$ [%]		
	−0·26	0	
	polarization:		
	$\Delta K_{\beta5}/K_{\beta1,3}$ [%]		$\varepsilon_{Cr III} = 1{\cdot}6$, $\varepsilon_{Cl} = 3{\cdot}1$ } ionizity $= 0{\cdot}43$
	2·2	2·1	
	total: $\Delta K_\beta/K_{\beta1,3}$ [%]		$3 \times 0{\cdot}43 = 1{\cdot}29$ ch. u. delocalized
	1·9	2·1	
	$R_{\beta\alpha}$ (related to Scofield)		
0·1359 ± 0·0002	0·136	0·137	0·136

at the very right hand side of Tab. 3 for the chosen cases of chromate, Cr VI, and chromiumchloride, Cr III, which may be compared with experiment and the free-atomic situation as references. More detailed and realistic information is, of course, provided by a multicentre approach, introduced also in this connection by *E. Hartmann* and accepted by *Band* et al. [33]. The middle and the left hand parts of the table prove both versions rather consistently, but their difference from the experiments cannot be ignored: they tend to show larger effects. This is the more surprising, since the following mechanism is disregarded: due to the EC process a neutrino emission and thus a nuclear recoil takes place (energy transfer in the order of 10 eV that means more than the dissociation energy), and this should result in a further and noticeable "polarization-like" deformation of the electron cloud within the 1s vacancy life-time, because the core orbitals will follow the recoil movement in contrast to the molecular orbitals, the latter being "glued" to the ligands. Thus a further K_β and a $R_{\beta\alpha}$ enhancement of some percent is expected [34].

In spite of a far-reaching theoretical effort, which by the help of its relatively realistic cluster approach revealed, in a similar way as in the case of nuclear decay rate variations, the special role of polarization contributions due to the ligand field, precise radiometry, as far as it is pursued as an experimental, faultless and thus authoritative guide, requires further efforts for the interpretation of chemical effects. These effects are small, indeed, but perhaps they carry some new aspects into the exceedingly successful working field which, however, has not experienced any great enrichment since its foundation in the thirtees and its "classical period" in the sixtees, as far as the establishment of really new methods is concerned; and they give rise to an interesting feedback to the arsenal of methods and even knowledge in atomic physics.

Acknowledgement

The author is indebted to all his collegues of the former "nuclear decay"-group for successful cooperation over years, to *H.-J. Thomas* and *A. Freyer* for new experimental contributions on $R_{\beta\alpha}$ and to Dr. *E. Hartmann* for his theoretical commitment and special discussions.

Received March 19, 1986

References

[1] A Handbook of Radioactivity Measurements Procedures (NCRP Report no. 58), Bethesda: Nat. Council on Rad. Prot. and Meas. 1985, p. 1

[2] *G. Brunner*, J. Phys. B: At. Mol. Phys. **17** (1984) 2971

[3] B. A. Кравцов, массы атомов и энергии связи ядер, атомиздат 1974

[4] *G. Brunner*, „Radioaktivität und das Problem der Zeit", ZfI-Bericht Leipzig 1974 (unveröff.)

[5] *P. A. M. Dirac*, Nature **139** (1937) 323; see also in Proc. Roy. Soc. A **165** (1938) 199

[6] *G. Brunner*, Hinweise auf eine zeitliche Änderung von Kernzerfallswahrscheinlichkeiten, ZfI-Bericht Leipzig 1974 (unveröff.)

[7] *E. K. Gerling, G. V. Ovchinnikova, O. A. Levchencov*, On the interpretation of the discordant apparent ages of the lunar and earth rocks and meteorites, Soviet-American Conf. on Cosmochem. of the Moon and Planets, June 4–8, 1974, Moscow, Abstracts p. 93 (paper no. 4.8)

[8] *H. J. Kreuzer, M. Gies, G. L. Malli, J. Ladic*, J. Phys. A: Math. Gen. **18** (1985) 1571

[9] *B. Crasemann*, Nucl. Instr. Meth. **112** (1973) 33

[10] *K.-P. Dostal, M. Nagel, D. Pabst*, Z. Naturforsch. **32 a** (1977) 345

[11] *R. Der, D. Hinneburg, M. Nagel*, Comput. Phys. Commun. **18** (1979) 401

[12] *H. C. Pauli, U. Raff*, Comput. Phys. Commun. **9** (1975) 392

[13] *V. A. Krutov, V. N. Fomenko*, Ann. Phys. (Leipzig) **21** (1968) 291

[14] *D. Hinneburg, M. Nagel, G. Brunner*, Z. Phys. A – Atoms and Nuclei **299** (1981) 201

[15] *D. Hinneburg, M. Nagel, G. Brunner*, Z. Phys. A **291** (1979) 113

[16] *J. C. Slater*, Quantum Theory of Molecules and Solids, McGraw-Hill, 1974

[17] *E. Hartmann, R. Der, M. Nagel*, Z. Phys. A **290** (1979) 349

[18] *E. Hartmann, G. Seifert*, phys. stat. sol. (b) **100** (1980) 589

[19] *M. Nagel, G. Brunner, K.-P. Dostal, E. Hartmann, D. Hinneburg, H. Kupsch, D. Pabst, A. Rigo*, Z. Naturforsch. **33a** (1978) 1050

[20] *K.-P. Dostal, B. Bayerl*, Z. Naturforsch. **35a** (1980) 894

[21] *J. C. Slater*, Phys. Rev. **84** (1951) 1261

[22] *K. T. Bainbridge, M. Goldhaber*, Phys. Rev. **84** (1951) 1260

[23] *B. Johannsen, R. Münze, K.-P. Dostal, M. Nagel*, Radiochem. Radioanal. Lett. **47** (1981) 57, see also in much more detail: *R. Münze, J. Bechstein, B. Großmann, E. Hartmann, J. Hoffmann, B. Johannsen, D. Mocker, B. Noll, F. Seifert, H. Spiess, R. Syhre*, G4-Research Report 1975–81, ZfK-RI 1981

[24] *O. Dragoun*, Int. Conf. on X-Ray and Inner-shell Processes in Atoms, Molec. and Solids (X-84), 20–24 Aug. 1984, Leipzig, Abstracts I, p. 123

[25] *M. Fiser, V. Brabec, O. Dragoun, A. Kovalik, J. Frana, M. Rysavy*, Int. J. Appl. Radiat. Isot. **36** (1985) 219

[26] *E. Hartmann, Ch. Eifrig, G. Brunner*, Hyperfine Interact. **12** (1982) 219

[27] *E. Hartmann*, Int. Conf. on X-Ray and Inner-shell Processes in Atoms, Molec. and Solids (X-84), 20–24 Aug. 1984, Leipzig, Proceedings, p. 297

[28] *K. Makarjunas*, Phys. Lett. **91A** (1982) 249

[29] *G. Brunner, A. Freyer, H.-J. Thomas*, ZfI-Mitt. **105** (1985) 63

[30] *E. Arndt, G. Brunner, E. Hartmann*, J. Phys. B: At. Mol. Phys. **15** (1982) L 887

[31] *G. Brunner, M. Nagel, E. Hartmann, E. Arndt*, J. Phys. B: At. Mol. Phys. **15** (1982) 4517

[32] *L. Pauling*, The Nature of the Chemical Bond, Ithaka, New York: Cornell Univ. 1960, p. 94

[33] *I. M. Band, A. P. Kovtun, M. A. Listengarten, M. B. Trzhaskovskaya*, J. Electron Spectrosc. Rel. Phenomena **36** (1985) 59

[34] *G. Brunner*, J. Phys. B: At. Mol. Phys. **16** (1983) 1199

Isotopenpraxis 22 (1986) 12, S. 425 bis 429

Beiträge der Radiotracertechnik zur Rationalisierung einer großtechnischen Anlage zur Caprolactamproduktion

R. Otto, P. Hecht, H.-G. Könnecke, H. Koch

(Akademie der Wissenschaften der DDR, Zentralinstitut für Isotopen- und Strahlenforschung, Leipzig)[1])

H. J. Naumann, W. Winzer, H. Grundmann

(VEB Leuna-Werke „Walter Ulbricht", Leuna)[2])

Herrn Prof. Dr. rer. nat. habil. *Klaus Fuchs* zum 75. Geburtstag gewidmet

Es wird über den Einsatz der Radiotracertechnik zur Erhöhung der Effektivität einer Caprolactamproduktionsanlage berichtet. Verweilzeitmessungen und Stoffverteilungsuntersuchungen wurden in den Verfahrensstufen Cumenoxidation, Phenolteerspaltung, Phenolhydrierung, Lactamextraktion und Caprolactamdestillation mit Hilfe von Natrium-24, Brom-82 und Technetium-99m durchgeführt. Die Versuchsergebnisse führten zu wesentlichen Durchsatzerhöhungen der Anlage, zur Verbesserung der Material- und Energieökonomie und zur Erhöhung der Qualität des Endproduktes Caprolactam. Insgesamt betrug das Aufwand-Nutzen-Verhältnis 1:12. Aus den problemorientierten Aufgabenstellungen ergaben sich methodische Weiterentwicklungen zur Markierungs-, Meß- und Auswertetechnik.

This paper reviews ways in which radioactive tracer techniques can be used to improve the efficiency of a caprolactam production plant. Residence time spectra and material distributions in different parts of the plant were measured by application of sodium 24, bromine 82 and technetium 99m. The investigations led to improved cumeneoxidators and gave useful informations on the catalytic decomposition of the phenolic tar and the hydrogenation process of phenol. It could be shown that the construction of the lactam extractor must be varied and the results of the tracerexperiments made in the distillation unit contributed to a higher quality of the final product.

Keywords

bromine 82; chemical industry; efficiency; lactams; optimization; phenol; production; sodium 24; technetium 99; tracer techniques

1. Einleitung

Die Intensivierung chemischer Produktionsanlagen ist ein Weg zur Erhöhung der Effektivität der chemischen Industrie. Bestehende Produktionsanlagen müssen effektiver ausgelastet, der spezifische Material- und Energieverbrauch muß gesenkt und eine höhere Qualität der Finalprodukte erreicht werden. Die zur Ableitung von konkreten Intensivierungsmaßnahmen notwendigen Prozeßanalysen sind dabei vorwiegend bei laufender Produktion durchzuführen. Gerade unter dieser Forderung kann die Radionuklidtracertechnik wesentliche Beiträge liefern.

In den VEB Leuna-Werken wird das ε-Caprolactam nach dem klassischen Verfahren [1] über folgende Stufen hergestellt:

- Alkylierung von Benzen mit Propen zu Cumen,
- Oxidation des Cumens mit Luft und säurekatalytische Spaltung des entstandenen Cumenhydroperoxids zu Phenol und Aceton,
- selektive Hydrierung des Phenols zu Cyclohexanon,
- Oximierung des Cyclohexanons, Umlagerung des Oxims mit Oleum und Neutralisation des Reaktionsproduktes, Abtrennung der öligen Caprolactamphase,
- Extraktion des Lactamöls mit Trichlorethen und nachfolgende Reextraktion des Caprolactams mit Wasser,
- Destillation des Rohlactams.

[1]) Anschriften: DDR-7050 Leipzig, Permoserstraße 15
[2]) DDR-4220 Leuna 3

Seit Inbetriebnahme der Anlage wurden in allen Verfahrensstufen Forschungs- und Entwicklungsarbeiten zur Effektivitätssteigerung durchgeführt [2]. Radiotracertechnische Untersuchungen erfolgten in den Prozeßstufen (Abb. 1):

- Cumenoxidation,
- Phenolteerspaltung,
- Phenolhydrierung,
- Reextraktion des Caprolactams,
- Destillation des Rohlactams.

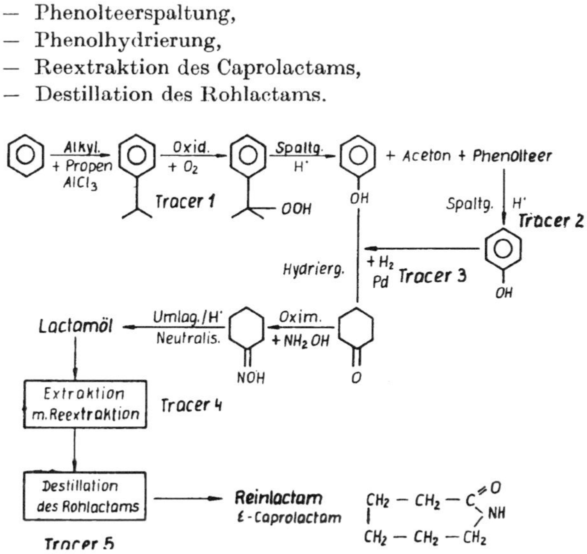

Abb. 1. Tracerexperimente in der Anlage zur Produktion von Caprolactam

2. Verweilzeitmessungen in der Verfahrensstufe Cumenoxidation

Die Anlage zur Luftoxidation des Cumens besteht aus parallel geschalteten kolonnenartigen Oxidatoren. An den Kolonnenböden wird Luft und an den Kolonnenköpfen Cumen zugeführt. Das Cumen enthält einen geringen Anteil an Cumenhydroperoxid (CHP) als Katalysator. Im Verlauf der Reaktion wird aufoxidiert, anschließend das CHP destillativ angereichert und das Rückcumen mit Frischcumen erneut der Oxidation unterworfen. In der bestehenden Anlage konnte jedoch eine geplante Durchsatzerhöhung nicht verwirklicht werden, da mit den eingebauten Glockenböden bei erhöhter Belastung kein stabiles Betriebsregime zu erreichen war. Durch Verweilzeitmessungen mit Radionukliden wurden die Ursachen für die dynamischen Instabilitäten aufgeklärt. Nach dem Passieren des Vorheizers setzte man dem Einsatzprodukt ^{82}Br-markiertes Bromnaphthalin zu. An allen Böden entnahm man Produktproben und bestimmte deren spezifische Zählraten. Die Verweilzeitspektren sind in der Abb. 2 dargestellt. Die mittleren Verweilzeiten auf den Böden enthält die Tab. 1. Die mittleren Verweilzeiten variieren sehr stark. Die längste Verweilzeit auf dem obersten Boden wurde erwartet, die sehr kleinen Verweilzeiten auf den Böden 8 und 6 beweisen ein „Durchfallen" des Produktes. Unmittelbar nach der Übergabe der Ergebnisse wurden die Oxidatoren bei geringem technischen Aufwand mit effektiveren Böden versehen. Jetzt wurde ein stabiles Betriebsregime auch bei gesteigerter Cumen- und Luftbelastung erreicht. Nach Realisierung der technischen Änderungen an den Oxidatoren auf der Basis der Tracerexperimente konnte der Durchsatz wesentlich gesteigert werden.

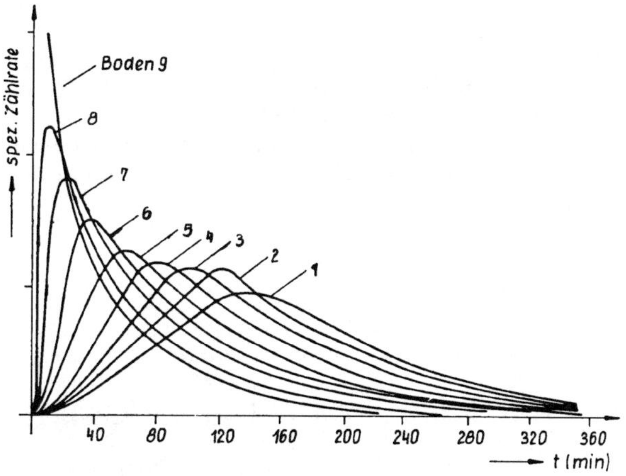

Abb. 2. Verweilzeitspektren des Cumens im Oxidator (Tracer 1)

Tab. 1. Tracerexperiment am Oxidator

Boden	mittlere Gesamtverweilzeit [min]	mittlere Verweilzeit auf dem Boden [min]
9	69,8	69,8
8	71,6	1,8
7	92,2	20,6
6	94,3	2,1
5	104,9	10,6
4	115,5	10,6
3	137,3	21,8
2	150,5	13,2
1	172,2	21,7

3. Messung von Mitreißeffekten in der Verfahrensstufe Phenolteerspaltung

Bei der säurekatalytischen Spaltung des Cumenhydroperoxids zu Phenol und Aceton fällt als Nebenprodukt Phenolteer an, dessen rationelle Aufbereitung zu Phenol zur Verbesserung der Ökonomie des Gesamtprozesses beiträgt. Im Zuge der geplanten Durchsatzerhöhung der Gesamtanlage mußte die dadurch ansteigende Teermenge unter der Voraussetzung verarbeitet werden, daß auch jetzt keine hochsiedenden und Sumpfbestandteile vor allem als mitgerissene Tröpfchen ins Destillat gelangen. Zur Ermittlung eines dann eventuell auftretenden Mitreißens von Flüssigkeitströpfchen wurde die Radioindikatoranalyse eingesetzt.

Das Einsatzprodukt wurde im Vorratstank (Abb. 3) mit ^{24}NaCl als nichtverdampfenden aber löslichen Tracer markiert. Das Destillat kann nur dann Aktivität enthalten, wenn Einsatzprodukt durch Mitreißen übergeht. Von Destillat und Einsatzprodukt entnahm man Proben und ermittelte deren spezifische Zählraten. Sowohl während des Anfahrens der Anlage als auch während zweier verschiedener Betriebszustände, auch bei verdoppeltem Durchsatz, gelangten nur 0,1 – 0,4% des Einsatzproduktes ins Destillat (Tab. 2).

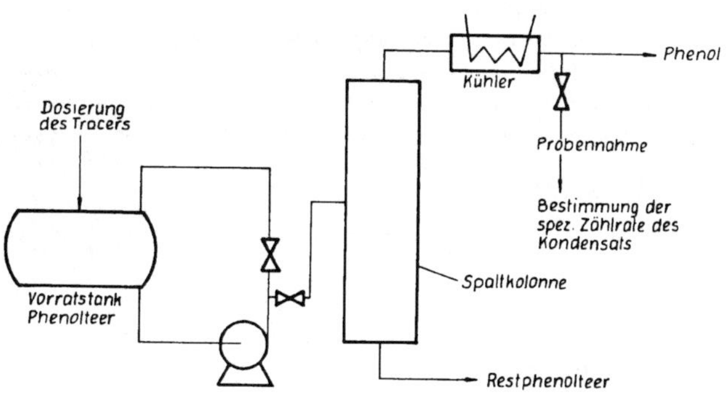

Abb. 3. Ermittlung von Mitreißeffekten in der Phenolteerspaltung (Tracer 2)

Tab. 2. Tracerexperiment in der Phenolteerspaltanlage

Anfahrzustand		Betriebszustand I		Betriebszustand II (verdoppelter Durchsatz)	
Proben-Nr.	G [%]	Proben-Nr.	G [%]	Proben-Nr.	G [%]
4	< 0,1	14	0,2	21	0,2
5	0,2	15	0,2	22	< 0,1
6	0,2	16	0,3	23	< 0,1
7	0,1	17	0,2	24	< 0,1
8	0,1	18	0,2	25	0,2
9	0,4	19	0,2	26	< 0,1

G ist der Anteil an mitgerissenem Einsatzprodukt im Destillat

Deshalb konnten die für eine Durchsatzerhöhung vorgesehenen technischen Veränderungen an der Anlage zur Vermeidung von vermuteten Mitreißeffekten sowie ein mit hohem technischen und analytischen Aufwand verbundener Großversuch mit anderen Füllkörpern entfallen. Die Ergebnisse der Tracerversuche bildeten hier die Grundlage für die wissenschaftlich-technische Absicherung der Durchsatzerhöhung in der Phenolteerspaltanlage bei gleichbleibender Qualität des gewonnenen Phenols.

4. Messung von Mitreißeffekten in der Anlage zur selektiven Hydrierung von Phenol [3]

In der Verfahrensstufe Selektivhydrierung wird Phenol an einem Palladiumkatalysator zu Cyclohexanon hydriert. Phenol führt man kontinuierlich einem Verdampfer zu, aus dem es bei erhöhter Temperatur von durchströmendem Wasserstoff ausgetragen wird (Abb. 4). Das austretende Phenoldampf-Wasserstoff-Gemisch wird einem Produktvorheizer zugeführt. Hier wird es auf die Reaktionstemperatur gebracht und gelangt nun in den Hydrierreaktor. Trotz der im oberen Teil des Verdampfers zur Abscheidung von mitgerissenen Phenoltröpfchen eingebauten Schicht aus Faltenkörpern trat im Produktvorheizer flüssiges Phenol auf, das Störungen im Betriebsablauf verursacht. Der Katalysator wird irreversibel geschädigt. Um mit geeigneten Maßnahmen das Auftreten von Flüssigphenol zu verhindern, mußte die Ursache seiner Bildung aufgeklärt werden. Die Abscheidung kann durch eine Kondensation in der Rohrleitung zum Vorheizer oder durch ungenügende Wirkung des Abscheiders entstehen. Die Ursachenermittlung erfolgte wieder mit Hilfe der Radioindikatoranalyse. Eine wäßrig-phenolische Lösung von $^{24}NaCl$ wurde in den Verdampfer dosiert und nach homogener Vermischung des Tracers mit dem Einsatzprodukt die spezifische Zählrate entnommener Einsatzprodukt- und Kondensatproben gemessen. Die Kondensatproben können wiederum nur Aktivität enthalten, wenn Flüssigkeitströpfchen mitgerissen werden. Untersucht wurden der Anfahrzustand und das normale Betriebsregime.

Die Auswertung der Ergebnisse ergab, daß lediglich zu Beginn des Anfahrzustandes ein geringer Mitreißeffekt von 0,4% signifikant nachweisbar war (Tab. 3).

Das Auftreten von Flüssigphenol im Produktvorheizer ist somit eindeutig auf eine Taupunktsunterschreitung in der Verbindungsleitung zurückzuführen. Nach zusätzlicher Beheizung und Isolierung der Verbindungsleitung trat keine Kondensatbildung mehr auf. Der Einbau eines zusätzlichen Demistors in der Anlage erübrigte sich. Ohne konstruktive Änderungen konnte nach den Ergebnissen der Traceruntersuchungen eine Mehrauslastung der Anlage bei gleichzeitiger Verlängerung der Lebensdauer des Katalysators erreicht werden.

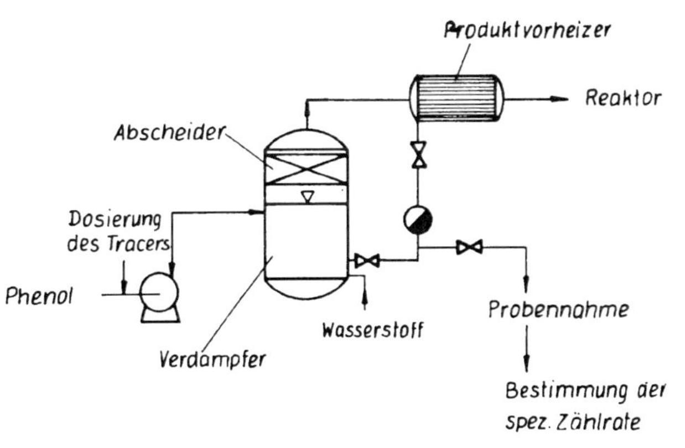

Abb. 4. Ermittlung von Mitreißeffekten in der Phenolhydrierung (Tracer 3)

Tab. 3. Tracerexperiment in der Phenolhydrierung

Anfahrzustand		Betriebszustand	
Proben-Nr.	G [%]	Proben-Nr.	G [%]
8	0,4	22	< 0,1
9	0,3	23	< 0,1
10	0,1	24	< 0,1
11	0,1	25	0,2
12	< 0,1	26	< 0,1
13	< 0,1	27	< 0,1

G ist der Anteil an mitgerissenem Einsatzprodukt im Kondensat

5. Verweilzeitmessungen in der Verfahrensstufe Reextraktion des Caprolactams

Nach der Oximierung des Cyclohexanons und der Umlagerung des Oxims zum ε-Caprolactam mit Oleum, sowie der anschließenden Neutralisation und der Abtrennung der öligen Caprolactamphase erfolgt die Extraktion des Caprolactams aus diesem sogenannten Lactamöl mit Trichlorethen. Daraus wird das Caprolactam zur weiteren Reinigung mit Wasser reextrahiert. Die Extraktion wird mit einem neuentwickelten Extraktortyp durchgeführt. Jedoch konnten die projektierten Caprolactamgehalte im zurückgeführten Trichlorethen und im Reextrakt nicht erreicht werden. Durch Ermittlung optimaler Betriebsparameter sollte der Caprolactamgehalt im rückgeführten Trichlorethen erniedrigt und im wäßrigen Reextrakt erhöht werden. Aus diesem Grunde wurde das dynamische Verhalten der Phasen und die Rückvermischung mit Hilfe von Radiotracern untersucht (Abb. 5). Die wäßrigen Phasen, Rein- und Brüdenkondensat, markierte man mit $^{24}NaCl$ und die organische Phase mit ^{82}Br-Brombenzen. An fünf ausgewählten Positionen des Schwingreextraktors (SRE) wurden in 13 Versuchen Verweilzeitspektren aufgenommen.

Die Tab. 4 enthält einige charakteristische Meßergebnisse für Rein- und Brüdenkondensat und die organische Phase. Es wurden wesentlich kürzere Verweilzeiten als erwartet gemessen. Am Reextraktorkopf sind die mittleren Verweilzeiten, wie erwartet, wesentlich größer als in den unteren Abschnitten des Reextraktors. Die Gesamtverweilzeit für die organische Phase ist größer als die für die wäßrige Phase.

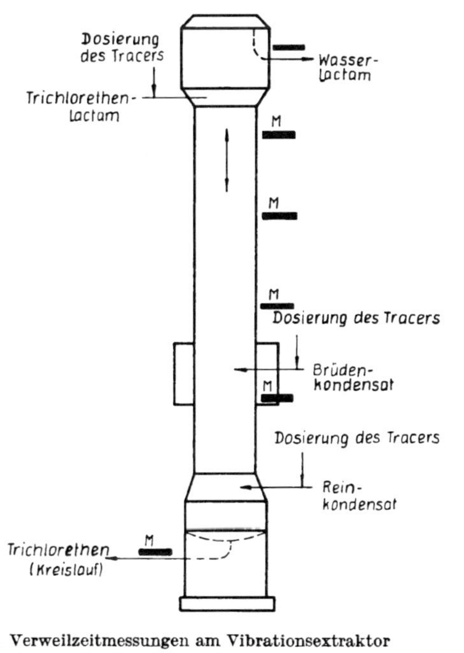

Abb. 5. Verweilzeitmessungen am Vibrationsextraktor (Tracer 4), (M: Meßstelle)

Tab. 4. Tracerexperiment in der Reextraktion

Phase	Abschnitt	L [m]	mittlere Verweilzeit [min]
Reinkondensat	M0 – M1	2,0	1,34
	M1 – M2	3,1	1,35
	M2 – M3	3,1	1,16
	M3 – M4	3,1	0,97
	M4 – M5	2,3	12,2
	M0 – M5	13,6	17,0
Brüdenkondensat	M0 – M2	2,3	2,34
	M2 – M3	3,1	0,70
	M3 – M4	3,1	1,65
	M4 – M5	2,3	13,40
	M0 – M5	10,9	18,09
Tri-Lactam	M0 – M1	13	10,1
	M1 – M2	3,1	5,02
	M2 – M3	3,1	6,07
	M3 – M4	3,1	6,70
	M4 – M5	4,9	16,4
	M0 – M5	15,5	44,3

Aus dem zeitlichen Verlauf der Intensitätsfunktion (Austrittswahrscheinlichkeit) ergeben sich deutliche Unterschiede im Verhalten der Phasen (Abb. 6). Ein Teilstrom der wäßrigen Phase wird infolge by-pass-Bildung, erkennbar am Maximum vor $t = \tau$ (τ = mittlere Verweilzeit), nicht oder nur ungenügend am Stoffaustausch beteiligt. Im oberen Teil des Reextraktors ist die Tendenz zur by-pass-Bildung geringer. Für die organische Phase hingegen zeigt die Intensitätsfunktion an allen Meßstellen das erwartete dynamische Verhalten. Durch die Tracerexperimente konnte daher bewiesen werden, daß höhere Caprolactamgehalte im Reextrakt nicht allein durch Änderung der Betriebsparameter zu erreichen sind. Die notwendige Produkt- und Energieökonomieverbesserung der Reextraktion kann nur durch konstruktive Veränderungen in der Extraktionskolonne erreicht werden, z. B. durch Veränderung der Eintrittsvorrichtung für die wäßrige Phase in den Extraktor, durch Veränderungen an den Kolonneneinbauten u. ä.

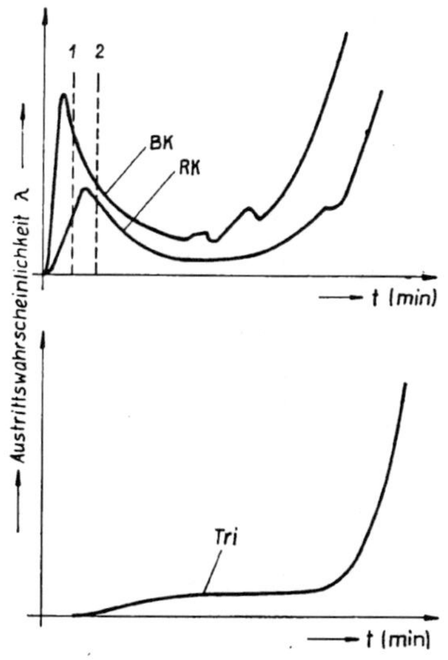

Abb. 6. Beispiele für den typischen zeitlichen Verlauf der Intensitätsfunktion für Brüdenkondensat (BK), Reinkondensat (RK) und Trichlorethen-Lactam (Tri)

(1: mittlere Verweilzeit des BK; 2: mittlere Verweilzeit des RK)

6. Verweilzeitmessung in der Verfahrensstufe Caprolactam-Destillation

Das aus der Wassereindampfungsstufe kommende Rohlactam wird abschließend in einer 3-stufigen Dünnschichtverdampfungsrotationsanlage zur Gewinnung des Reinlactams einer schonenden Destillation unter vermindertem Druck unterworfen (Abb. 7). In der ersten Stufe erfolgt die Entwässerung des Rohlactams, in der 2. Stufe die Abtrennung einer Caprolactamfraktion mit einem Anteil an leichtflüchtigen Verunreinigungen und in der dritten Stufe wird Reinlactam abdestilliert. Die Sumpfphasen werden entsprechend in die nächste Destillationsstufe gepumpt. Als Bestandteil der Prozeßanalyse waren Verweilzeitmessungen der nichtflüchtigen Sumpfprodukte in den Destillationsstufen 1 und 2 mit dem Ziel durchzuführen, das Sumpfprodukt-Destillat-Verhältnis aus energetischen Gründen zu optimieren und ein rechtzeitiges regelungstechnisches Eingreifen in den Prozeß bei Anfall qualitätsgeminderten Rohlactams zu ermöglichen. An der Druckseite der Dosierpumpe zur 1. Destillationsstufe wurde ^{99m}Tc-Pertechnetat injiziert. Nach den Dosierpumpen zur 2. und 3. Destillationsstufe entnahm man Sumpfproduktproben, bestimmte deren spezifische Zählraten und berechnete die mittleren Verweilzeiten. Die Ergebnisse sind in der Tab. 5 und der Abb. 8 dargestellt.

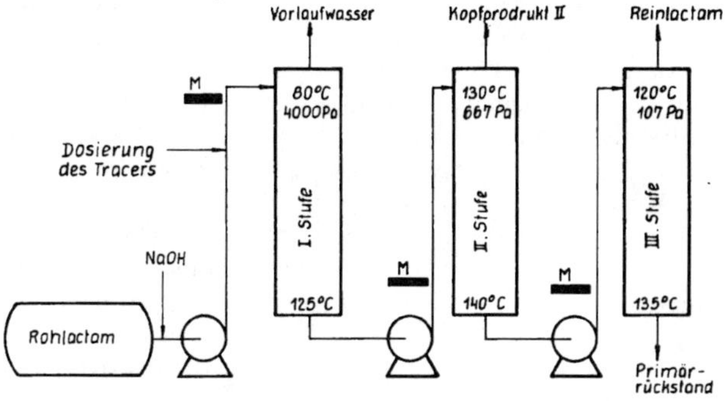

Abb. 7. Verweilzeitmessungen in der DSVR-Anlage (Tracer 5), (M: Meßstelle)

Tab. 5. Tracerexperiment in der DSVR-Anlage

Abschnitt	mittlere Verweilzeit [min]
Stufe 1	38,2
Stufe 2	41,2
Stufe 1 + 2	79,4

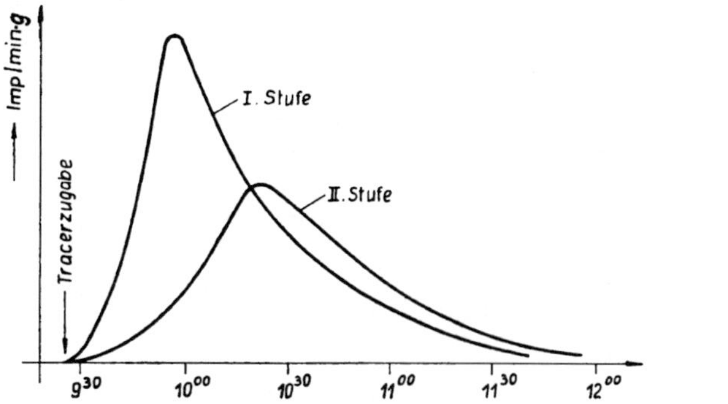

Abb. 8. Verweilzeitspektren des Sumpfproduktes in der I. und II. Destillationsstufe der DSVR-Anlage

Die mittleren Verweilzeiten der nichtverdampfbaren Sumpfprodukte liegen bei nur ca. 40 min und unterscheiden sich nicht signifikant voneinander. Entsprechend den Ergebnissen der Tracerexperimente wurde das Sumpfprodukt-Destillat-Verhältnis optimiert und es kann nun auch exakt die Zeit angegeben werden, nach deren Ablauf ein Einsatz von qualitätsgemindertem Rohlactam als Folge von Unregelmäßigkeiten in der Synthese eine mindere Qualität des Endproduktes verursacht bzw. man weiß jetzt, wann betriebliche Maßnahmen zur Aufrechterhaltung der vorgegebenen Qualität einzuleiten sind. Damit besteht der ökonomische Nutzen dieser Traceruntersuchungen, deren Ergebnisse einen wesentlichen Beitrag zum wissenschaftlichen Betriebsregime der Anlage lieferten, in der Sicherung einer hohen und gleichbleibenden Qualität des Endproduktes Reinlactam.

7. Einschätzung der Ergebnisse

Der in diesem Umfang durchgeführte Einsatz der Radionuklidtracertechnik in einem vielstufigen großtechnischen Produktionsprozeß der chemischen Industrie hat damit wesentlich

- zur Erarbeitung der wissenschaftlichen Grundlagen der einzelnen Verfahrensstufen,
- zur Optimierung der Betriebsregime,
- zur Erhöhung der Anlagenkapazität durch wissenschaftlich-technisch begründete konstruktive Veränderungen in den Anlagen bei möglichst geringem Aufwand,
- zur Qualitätsverbesserung des Finalproduktes,
- zur Einsparung von Forschungs- und Investmitteln beigetragen.

Der erzielte Nutzen ist umso höher zu bewerten, da die Überführung der Ergebnisse in die Praxis unmittelbar nach Übergabe derselben an den Auftraggeber erfolgte. Das Aufwand-Nutzen-Verhältnis insgesamt beträgt 1:12,1. Für die einzelnen Verfahrensstufen ergeben sich folgende Aufwand-Nutzen-Verhältnisse:

Cumenoxidation	1:30,7
Phenolteerspaltung	1:5,4
Phenolhydrierung	1:18,3
Reextraktion	1:1,5
Destillation	1:15,0

Die Radionuklidtracertechnik hat sich damit erneut als ein wesentlicher und nützlicher Bestandteil von Prozeßanalysen und einer gezielten Verfahrenspflege auch und insbesondere an vielstufigen großtechnischen Produktionsanlagen unter Betriebsbedingungen bewährt.

Die beschriebenen Experimente haben aber auch zur methodischen Weiterentwicklung der Tracertechnik beigetragen. Diese methodischen Arbeiten im Rahmen der verfahrenstechnischen Aufgabenstellungen wurden zur Sicherung des notwendigen methodischen Vorlaufs sehr zielbewußt betrieben.

An den folgenden Beispielen soll das verdeutlicht werden:

1. Um den speziellen Anforderungen der Markierung bei der Cumenoxidation zu genügen, wurden neue Methoden zur Einführung von ^{82}Br und ^{131}I in organische Verbindungen entwickelt. Sie beruhen u. a. auf speziellen Isotopenaustauschreaktionen. Die Methoden sind publiziert [4] und für die Markierung bei großtechnischen Experimenten auch in anderen Betrieben genutzt und erprobt worden.
2. Probleme der Markierung bei der Verweilzeitmessung in der Caprolactamdestillation führten zur Ausarbeitung von Methoden zur schnellen Überführung wäßriger ^{99m}Tc-, ^{113m}In- und ^{132}I-Eluate von Radionuklidgeneratoren in organische Phasen sowie zur Entwicklung eines einsatzfähigen ^{140}Ba/^{140}La- und ^{137}Cs/^{137m}Ba-Generators [4].
3. Um eine eindeutige Zuordnung von Meßwert und Meßzeitpunkt bei Verweilzeitmessungen mit Hilfe der vorhandenen Meßanordnung (Strahlenmeßgerät Typ 20026 und Ergebnisdrucker Typ 23144, beide vom VEB RFT Meßelektronik „Otto Schön“, Dresden) zu gewährleisten, wurde ein einfaches externes Steuergerät entwickelt, wodurch die Zähler nach beliebig einstellbaren Wiederholzeiten simultan gestartet werden [5].

Eingegangen am 16. April 1986

Literatur

[1] *W. Rösler, H. Lunkwitz*, Chem. Techn. **30** (1978) 67
[2] *H. J. Naumann*, Mitteilungsblatt der Chem. Ges. der DDR **26** (1979) 215, 234
[3] *R. Otto, H. G. Könnecke, P. Hecht, H. Ulrich*, ZfI-Mitt., **39** (1981) 55
[4] *R. Otto*, ZfI-Mitt. **109** (1985) 67
[5] *P. Hecht, R. Otto, H. Ulrich*, Isotopenpraxis **19** (1983) 3

Isotopenpraxis 22 (1986) 12, S. 430 bis 435

Ionenstrahlätzen — Verfahren und Ausrüstungen für die Mikrostrukturierung

F. Bigl

(Akademie der Wissenschaften der DDR, Zentralinstitut für Isotopen- und Strahlenforschung, Leipzig)[1])

Herrn Prof. Dr. rer. nat. habil. *Klaus Fuchs* zum 75. Geburtstag gewidmet

Bei der Entwicklung und Herstellung hochintegrierter Halbleiterschaltkreise nimmt die Bedeutung von Strahlverfahren für Stoffmodifizierung und -analytik mit Verkleinerung der Abmessungen der Einzelbauelemente ständig zu. Der Beitrag beschreibt Möglichkeiten und Probleme des Einsatzes von niederenergetischen Ionenstrahlen für Teilschritte der sog. „Strukturübertragung" und skizziert den im Zentralinstitut für Isotopen- und Strahlenforschung erreichten Entwicklungsstand.

Ion beam techniques are widely used for modification as well as analysis of materials in development and production of VLSI circuits and their importance is continuously increased in course of the reduction of structure dimensions. Some possibilities and problems connected with the application of ion beams of low energy in the pattern transfer are described and the stage of development achieved in the Central Institute of Isotope and Radiation Research is outlined.

Keywords

etching; ion beams; ion sources; surface finishing; semiconductor materials; silicon; silicon compounds; sputtering

1. Einleitung

Das Entwicklungstempo der Mikroelektronik war in den vergangenen 15 Jahren außergewöhnlich hoch. Die Entwicklung integrierter Schaltkreise war insbesondere durch die ständige Erhöhung des Integrationsgrades bzw. der Bauelemente-Packungsdichte und die damit verbundene Verringerung der Abmessungen der Einzelbauelemente, des Energiebedarfs und der Kosten pro Funktionselement sowie die Steigerung der Verarbeitungsgeschwindigkeit und der Zuverlässigkeit gekennzeichnet.

Mit dem unvermindert anhaltenden Trend zur Strukturverkleinerung eng verknüpft ist die Entwicklung und die Produktionseinführung neuer Herstellungstechnologien oder zumindest neuer Teilschrittverfahren und Spezialausrüstungen. Zunehmende Bedeutung erlangen dabei Verfahren und Geräte, die korpuskulare oder elektromagnetische Strahlung zur Modifizierung entsprechender Materialien, Schichten bzw. Festkörperoberflächen oder ihrer Untersuchung bezüglich Struktur oder Zusammensetzung nutzen (z.B. Elektronenstrahl-, Röntgen- und Ionenstrahllithografie, Ionenimplantation, Ionenstrahlbearbeitung, Laserbearbeitung, Strahl-Beschichtungsverfahren; AES, RBS, PIXE, SIMS, REM, TEM usw.).

Ein in Fertigungstechnologien von Halbleiterbauelementen häufig wiederkehrender Teilschritt ist die Strukturübertragung von Haftmasken (meist fotolithografisch strukturierte organische Resistschichten) in Halbleiter-, Metall- oder Isolatorschichten. Der sich gegenwärtig international vollziehende Übergang zu kleinsten Strukturbreiten um oder unter 1 µm erfordert dabei Übertragungsgenauigkeiten, die zunehmend von naß- oder plasmachemischen Ätzprozessen nicht mehr erfüllt werden können. An ihre Stelle treten hauptsächlich Gasphasenätzprozesse unter Ionenbeteiligung, die im allgemeinen einen hoch anisotropen Materialabtrag bewirken und dadurch Maskenunterätzungen vermeiden. Unter den verschiedenen Varianten dominieren gegenwärtig und in naher Zukunft Ionenätzprozesse in RF-Plasmaentladungen mit planaren bzw. zylindrischen Elektroden in Diodenanordnungen ohne oder mit Hilfsmagnetfeld in den Varianten IE (Ion Etching), RIE (Reactive Ion Etching), MIE (Magnetic confinement Ion Etching), MRIE (Magnetic confinement Reactive Ion Etching). Langfristig prognostizieren internationale Einschätzungen jedoch eine Zunahme des Anteiles von Ionenstrahlätzverfahren. In einer amerikanischen Studie wird ihr Anteil an Strukturübertragungsprozessen für 1988 auf 20% geschätzt [1]. Der folgende Beitrag zeigt an einigen Beispielen Probleme der Erforschung von Ionenstrahlätzprozessen und ihrer Nutzung für die Mikrostrukturierung. Gleichzeitig dokumentiert er den im Zentralinstitut für Isotopen- und Strahlenforschung der AdW und beim Ausrüstungsentwickler (VEB Hochvakuum Dresden) erreichten Entwicklungsstand.

2. Varianten des Ionenstrahlätzens und ihre charakteristischen Merkmale

Im Gegensatz zu Plasmaätzanordnungen sind in Ionenstrahlätzanlagen die Bereiche der Ionenerzeugung und ihrer Wechselwirkung mit dem Substrat deutlich räumlich getrennt. Die Probe befindet sich im feldfreien Raum und wird im wesentlichen nur von positiven Ionen (Energie im allgemeinen 0,3 ... 2 keV), die aus einem Niederdruckplasma (Arbeitsdruck ca. 10^{-2} Pa) über ein 2- oder 3-Gittersystem extrahiert werden, getroffen. Je nach Gasart (die im wesentlichen bestimmt, ob physikalische oder chemische Abtragungsprozesse dominieren) und Ort der Gaszuführung unterscheidet man drei Varianten des Ionenstrahlätzens:

IBE (Ion Beam Etching, oft auch IBM — Ion Beam Milling): Ein Inertgas (z.B. Argon) wird direkt in den Ent-

---

[1]) Anschrift: DDR-7050 Leipzig, Permoserstr. 15

ladungsraum der Ionenquelle eingelassen. Der Materialabtrag erfolgt durch physikalische Zerstäubung. Hoher Anisotropie und universellen Anwendungsmöglichkeiten für alle vakuumbeständigen Materialien stehen relativ niedrige Ätzraten und geringe Ätzratenunterschiede gegenüber. Der praktische Einsatz erfolgt insbesondere für in-situ-Oberflächenreinigungen vor Beschichtungsprozessen und zur Strukturierung relativ dünner, meist metallischer Schichten bzw. Schichtfolgen.

RIBE (Reactive Ion Beam Etching): Die Ionenquelle wird mit einem Gas bzw. Gasgemisch gespeist, das mit dem Material der zu ätzenden Schicht unter den Bedingungen des Ionenbeschusses chemisch reagieren und flüchtige Verbindungen bilden kann. Bei vergleichbaren Ionenstromdichten liegen die Ätzraten in der Regel 3 bis 5 mal höher als bei Zerstäubungsprozessen; durch geeignete Wahl von Gaszusammensetzung und Prozeßführung kann eine gute Selektivität gegenüber Maskenmaterial und Unterlage erreicht werden. Besonders bei höheren Ionenenergien kann der reaktive Materialabtrag von Materialzerstäubung begleitet werden.

CAIBE (Chemically Assisted Ion Beam Etching; identisch mit IBAE — Ion Beam Assisted Etching): Die Ionenquelle wird im allgemeinen mit einem Inertgas betrieben, die reaktive Gaskomponente wird direkt auf die Probenoberfläche geleitet. Bei ausreichender vakuumtechnischer Trennung von Entladungsraum und Bestrahlungsgebiet gestattet diese Variante in weiten Grenzen eine unabhängige Regelung von Ionen- und Neutralteilchenfluß — ein Vorteil, von dem in Zukunft vor allem eine bessere Steuerbarkeit von Kantenprofilen erwartet wird.

3. Ätzraten und ihre Abhängigkeit von Gasart und Prozeßparametern

Tab. 1 zeigt eine Zusammenstellung von (R)IBE- bzw. CAIBE-Abtragungsraten für ausgewählte Materialien und Ätzgase. Angegeben wurden auf eine Stromdichte von 1 mA cm^{-2} normierte mittlere Werte für senkrechten Strahleinfall und Ionenenergien von 0,5 bzw. 1,0 keV.

Für eine bestimmte Ätzgas-Material-Kombination ist die Ätzgeschwindigkeit im wesentlichen eine Funktion von Ionenstromdichte, Ionenenergie und Ioneneinfallswinkel. Der Einfluß von Gasdruck und Substrattemperatur ist i. a. relativ gering und kann bei den meisten Anwendungen vernachlässigt werden.

Abb. 1 dokumentiert den typischen Verlauf der Ätzraten — Energieabhängigkeit für physikalische und reaktive Materialabtragprozesse am Beispiel Ar/SiO_2 bzw. CHF_3/SiO_2. Während im betrachteten Energiebereich die Zerstäubungsrate nahezu linear mit der Ionenenergie ansteigt, erreicht der chemische Umsatz bereits zwischen 300 und 500 eV einen Sättigungswert [2]. Abb. 2 zeigt am Beispiel des Ätzens von Si mit C_2F_6 (RIBE bei Ionenenergien zwischen 0,6 und 1,8 keV) die Abhängigkeit der Ätzrate vom Ioneneinfallswinkel [3]. Der Verlauf weist darauf hin, daß bei niedrigen Ionenenergien der chemische Abtrag überwiegt

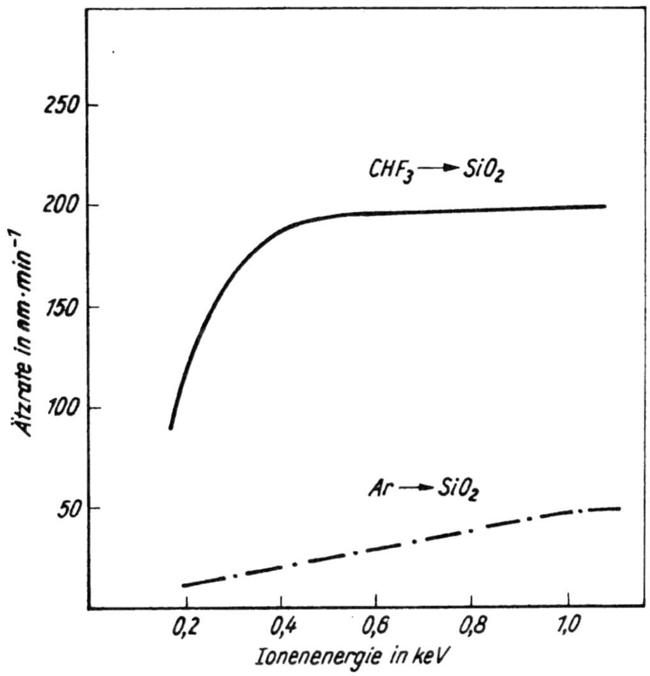

Abb. 1. Ätzraten von SiO_2 in Abhängigkeit von der Ionenenergie (normiert auf eine Stromdichte von 1 mA cm^{-2}) für das Ionenstrahlätzen mit Ar und CHF_3 (Arbeitsdruck 6 × 10^{-2} Pa) [2]

Tab. 1. Ätzraten in Angström (Å) je Sekunde nach mittleren Literaturwerten (normiert auf eine Stromdichte von 1 mA cm^{-2} und senkrechtem Ioneneinfall; — ≙ Ionenenergie 0,5 keV; + ≙ Ionenenergie 1 keV; ∘ = CAIBE bei 1 keV)

Material	Ätzgas								
	Ar	CHF_3	C_2F_6	CF_4	SF_6	Cl_2	Cl_2/Ar	CCl_4	CCl_4/Ar	N_2O
Poly-Si	13 +	2 —	3 —	7 —	50 +			40 —	50 +	
SiO_2	7 +	40 —	40 —	40 —	50 +			20 —	25 +	2 —
Si_3N_4	7 +		30 —	18 —				40 —		
Al_2O_3	2 +									
GaAs	36 +	15 —	15 —			50 +	50 ∘			
GaP	23 —									
InP	33 +							25 —		
$LiNbO_3$	8 —									
AZ 1350	10 +	2 —	10 —	15 —					20 —	50 —
PMMA	6 —	12 —	30 —	50 —						60 —
Al	20 +	12 —		4 —				25 —	28 —	
Ti	12 +						15 ∘			
Mo	11 +						17 ∘			
Au	42 +									
Ta	10 +									
Cu	18 +									
Cr	9 —									
Ni	10 —									
W	6 —									

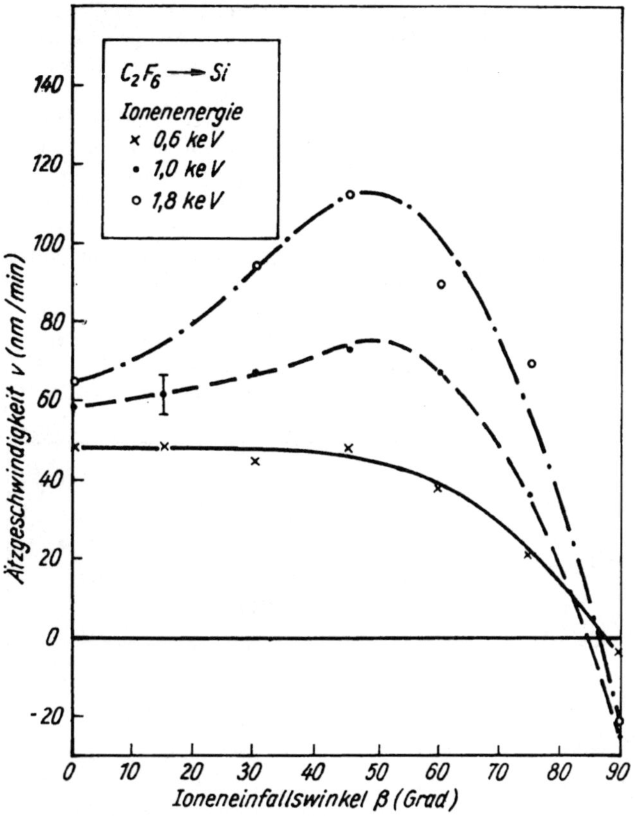

Abb. 2. Abhängigkeit der Si-Ätzgeschwindigkeit (RIBE mit C_2F_6 als Ätzgas) vom Ioneneinfallswinkel für Ionenenergien von 0,6, 1,0 und 1,8 keV (Ätzraten normiert auf 1 mA cm^{-2}) [3]

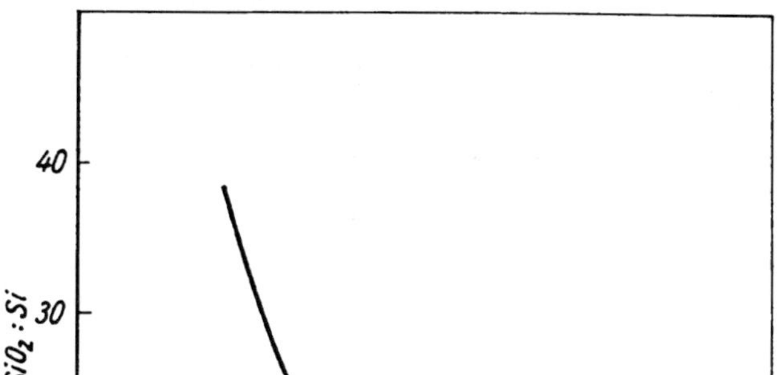

Abb. 3. Selektivität (Verhältnis der Ätzraten) SiO_2:Si in Abhängigkeit von der Ionenenergie für die Ätzgase CHF_3, C_2F_6 und CF_4 (Variante RIBE)

(Produktbildung an der Oberfläche weitgehend unabhängig vom Ioneneinfallswinkel, Ätzratenabfall durch Stromdichteverringerung in der Substratebene), bei höheren Energien dagegen der Zerstäubungseffekt dominiert (typischer Ätzratenanstieg bis zu Einfallswinkeln zwischen 40 und 60 Grad, danach steiler Abfall). Abb. 3 verdeutlicht, daß die sog. „Selektivität" (im Beispiel das Ätzratenverhältnis SiO_2/Si) sowohl vom Ätzgas als auch von der Ionenenergie abhängt. Im gezeigten Beispiel ist die hohe Selektivität bei niedrigen Ionenenergien und CHF_3 als Ätzgas auf eine Blockade der Si-Oberfläche durch Kohlenstoff oder abgeschiedene Polymerschichten zurückzuführen.

4. Strahlenschäden und Verunreinigungen nach Strahlätzprozessen

Ionenstrahlätzprozesse können unerwünschte Folgen haben:

- Die Struktur kristalliner Oberflächenbereiche wird durch den Ionenbeschuß stark gestört. Die Dicke der gestörten Schicht liegt i. a. noch über der Reichweite doppelt geladener Ionen des benutzten Ätzgases;
- In den oberflächennahen Bereich werden alle Elemente „implantiert", die sich im Ätz- und Restgas befinden oder während des Ätzprozesses durch Plasma- oder Strahlwechselwirkung mit Anlagenbauteilen in den Ionisationsraum oder auf die Probenoberfläche gelangen.

Untersuchungen zum Nachweis der Art und Tiefenverteilung von Strahlenschäden und Verunreinigungen sowie die Entwicklung von Methoden und Verfahren zu ihrer Minderung oder nachträglichen Beseitigung sind ein wesentlicher Bestandteil entsprechender Forschungsarbeiten [4]. Wegen der geringen Eindringtiefen von wenigen 10 nm ist besonders die Bestimmung von Tiefenprofilen relativ schwierig. Abb. 4 zeigt als Beispiel die Defekt-Tiefenverteilung in mit Ar^+-Ionen (Dosis 10^{17} cm^{-2}; Energie 0,7, 1,0, 1,6 und 2,0 keV) vorbehandeltem GaAs im Vergleich zum nichtbestrahlten Material (Defektdekoration mit Cu-64 und autoradiographischer Auswertung an 0,3-keV-Ionenstrahlschrägschliffen) [5].

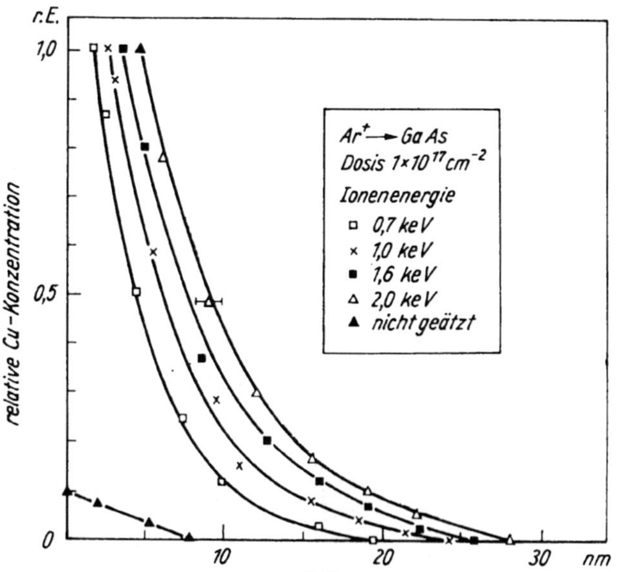

Abb. 4. Defekt-Tiefenverteilung in mit Ar-Ionen (Dosis 1×10^{17} cm^{-2}, Energie 0,7, 1,0, 1,6 und 2,0 keV) vorbehandeltem GaAs im Vergleich zum nicht bestrahlten Material (Defekt-Dekoration mit Cu-64 und autoradiographischer Auswertung an 0,3-keV-Ionenstrahlschrägschliffen) [5]

5. Strukturübertragung durch Ionenstrahlätzen

Wie bereits eingangs erwähnt, wird die Strukturübertragung von foto- oder elektronenstrahllithographisch hergestellten Haftmasken in halbleitertechnologisch relevante Schichten bei Strukturgrößen im Submikrometerbereich als eines der Hauptanwendungsgebiete der Ionenstrahlätztechnik erwartet. Aus Platzgründen kann im Rahmen dieses Beitrages nur auf die wesentlichsten Zielstellungen solcher Übertragungsprozesse hingewiesen werden. Abb. 5 zeigt das Ergebnis einer Rechnersimulation der Evolution bzw. Erosion von Kantenprofilen beim reaktiven Ionenstrahlätzen [6]. Die „Höhenlinien" verdeutlichen den von oben nach unten zeitlich fortschreitenden Profilverlauf im Maskenmaterial (I), in der zu ätzenden Schicht (II) und in der Unterlage (III). Die Hauptziele der Untersuchungen bestehen darin, bei vorgegebenem Masken-, Schicht- und Substratmaterial sowie vorgegebener Dicke der zu ätzenden Schicht die Maskendicke, das Maskenausgangsprofil und die Prozeßparameter (Gasart, Ionenenergie, Ioneneinfallswinkel usw.) so zu wählen, daß die Kantenverschiebung (KV) möglichst gering ist (im angenommenen Beispiel ca. 0,3 μm; bei tatsächlichen RIBE-Prozessen im allgemeinen unter 0,1 Mikrometer erreichbar), das Endprofil der Schichtkante eine gewünschte Form hat (Neigungswinkel im Beispiel ca. 76°) und notwendige Überätzungen (Inhomogenitäten in der Schichtdicke und der Strahlstromdichte!) die Unterlage möglichst wenig angreifen. Als Beispiel für das Ergebnis einer RIBE-Strukturübertragung zeigt Abb. 6 die rasterlektronenmikroskopische Aufnahme von Strukturen in einer 800 nm dicken SiO_2-Schicht (Ätzgas CHF_3, Ionenenergie 1 keV).

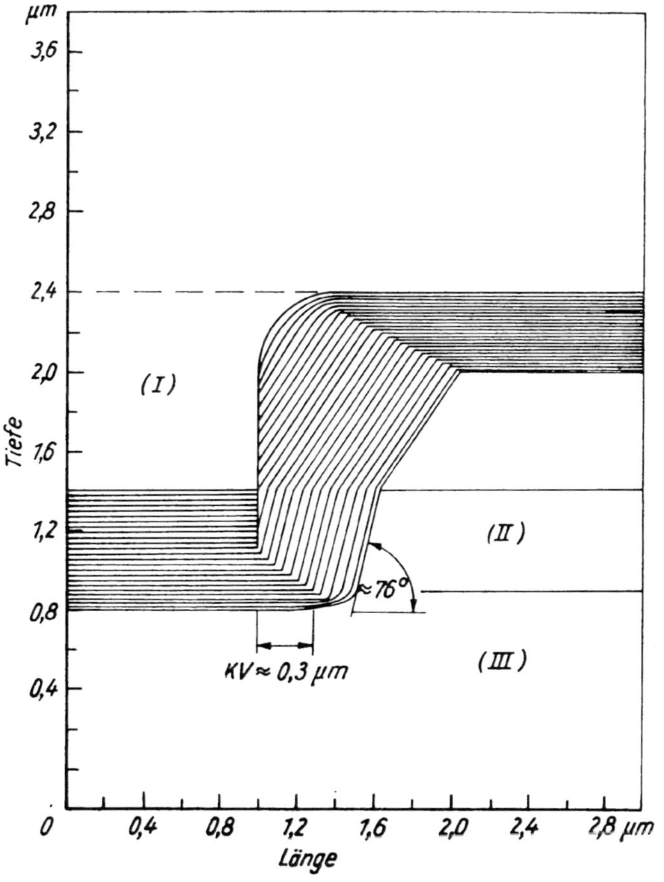

Abb. 5. Rechnersimulation der Evolution bzw. Erosion von Kantenprofilen bei der RIBE-Strukturübertragung von einer Haftmaske (I) in eine Schicht (II), die sich auf einer Unterlage (III) befindet [6]

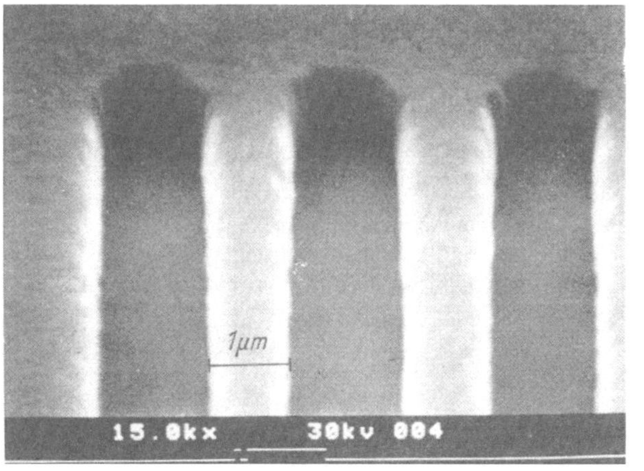

Abb. 6. Rasterelektronenmikroskopische Aufnahme von Strukturen in einer 800 nm dicken SiO_2-Schicht (RIBE mit CHF_3 bei 1 keV)

6. Ionenstrahl — „Schrägschlifftechnik"

Als weiteres Beispiel für die außerordentlich hohe Präzision des „Werkzeuges Ionenstrahl" bei der Bearbeitung von Festkörperoberflächen soll auf das im ZfI entwickelte Ionenstrahlschrägschliffverfahren hingewiesen werden [7], [8].

Bei einer geeigneten geradlinigen Relativbewegung zwischen einer ebenen Probe und einer Blende unter einem großflächigen Ionenstrahl läßt sich die Probenoberfläche unter extrem flachen Winkeln „anschleifen". Bei geeigneten Parametern lassen sich Neigungswinkel bis zu wenigen zehntausendstel Grad erreichen. Abb. 7 enthält Beispiele für Schrägschliffe in einer SiO_2-Schicht bzw. in einer Poly-Si/SiO_2/Si-Schichtfolge (Bestimmung der Tiefenprofile mit dem Gerät TALYSTEP). Durch Anwendung der neuen Präparationsmethode konnten in den letzten Jahren auch

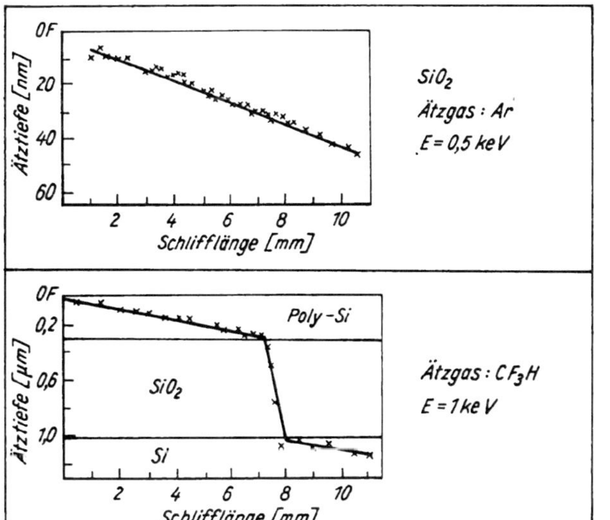

Abb. 7. Tiefenprofile von Ionenstrahlschrägschliffen in SiO_2 (Ätzgas Ar) und in einer Poly-Si/SiO_2/Si-Schichtfolge (Ätzgas CHF_3)

solche analytischen Verfahren für die Tiefenprofilanalyse eingesetzt werden, deren Tiefenauflösung ohne die extreme Profilstreckung nicht ausreicht (z. B. autoradiographische Verfahren).

7. Stand bei der Entwicklung von Ionenstrahlätzausrüstungen in der DDR

Vom VEB Hochvakuum Dresden und dem Zentralinstitut für Isotopen- und Strahlenforschung Leipzig wurden zwei Typen von Breitstrahlionenquellen mit nutzbaren Strahl-

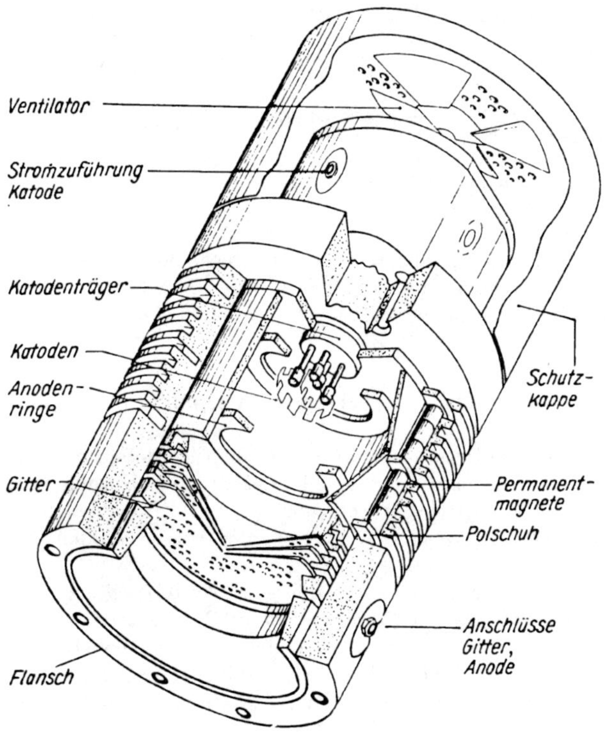

Abb. 8. Schnittdarstellung der Ionenquelle isq 76 (Hersteller: VEB Hochvakuum Dresden) [11]

Abb. 10. Ionenstrahlätzanlage ISA 150 (Hersteller: VEB Hochvakuum Dresden); Foto: Günther

durchmessern von 76 bzw. 150 mm (Strahlstromdichteabweichungen $\leq 5\%$ vom Mittelwert) mit den Typenbezeichnungen isq 76 und isq 150 und eine mikrorechnergesteuerte Ionenstrahlätzanlage ISA 150 entwickelt [9]. Die Abb. 8 und 9 zeigen eine Schnittdarstellung bzw. ein Foto der Ionenquelle isq 76, Abb. 10 die Gesamtansicht der Ionenstrahlätzanlage ISA 150. Betriebs- und Leistungsparameter der Ausrüstungen können Prospektmaterial entnommen werden, das der VEB Hochvakuum Dresden auf Wunsch zur Verfügung stellt [10], [11].

Eingegangen am 23. 3. 1986

Literatur

[1] Studie „Technologies for VLSI Dry Processes“, zit. in Solid State Technologies 27 (1984) 4

[2] *G. Ringel, F. Bigl, G. Wolf*, Gemeinsamer Jahresbericht ZfK-503 (1983) 96

[3] *G. Ringel, F. Bigl, H. Schmidt*, Gemeinsamer Jahresbericht ZfK-559 (1985) 126

Abb. 9. Breitstrahlionenquelle isq 76 (Hersteller: VEB Hochvakuum Dresden); Foto: Günther

[4] *D. Flamm, F. Bigl, W. Frank, W. Beyer, T. Lehmann, A. Dittmar,* Gemeinsamer Jahresbericht ZfK-559 (1985) 128
[5] *R. Fechner, A. Schindler, F. Bigl, J. Flachowsky,* Gemeinsamer Jahresbericht ZfK (in Vorbereitung)
[6] *G. Ringel, A. Nickel, F. Bigl,* Gemeinsamer Jahresbericht ZfK (in Vorbereitung)
[7] *F. Bigl, A. Schindler, R. Fechner, K. Freyer, H.-J. Treutler,* Proc. Int. Conf. on Ion Beam Modification of Materials, Budapest 1978, 1743
[8] *F. Bigl, A. Schindler, R. Fechner,* Proc. Sympos. Phys. Grundlagen zu Bauelementetechnologien der Mikroelektronik, Frankfurt/Oder 1979, 562
[9] *R. Schlenk, F. Kadner, K. Ehrig, F.Bigl, B. Faust, H. Neumann, A. Schindler,* Gemeinsamer Jahresbericht ZfK-530 (1984), 77
[10] „Ionenstrahlätzanlage ISA 150", Prospekt des VEB Hochvakuum Dresden
[11] „Ionenstrahlquellen isq 76 und isq 150", Prospekt des VEB Hochvakuum Dresden

Isotopenpraxis 22 (1986) 12, S. 435 bis 438

Catalytic Activity of $Ni-Cr_2O_3$ in the H_2-NH_3 and H_2-H_2O Deuterium Exchange Reactions

N. Palibroda, E. Grecu, P. Mărginean

(State Committee for Nuclear Energy, Institute of Isotopic and Molecular Technology)[1])

The activity of $Ni-Cr_2O_3$ catalyst for the deuterium exchange reactions between hydrogen and ammonia, as well as for hydrogen and water vapor has been measured in dependence on the reaction temperature and on partial pressure of ammonia and water vapor, respectively. In both cases the activity in dependence on the partial pressure shows a maximum; the maximum of activity for H_2-NH_3 exchange is situated between partial pressures of 0·05 and 0·25, and for H_2-H_2O reaction between 0·25 and 0·5. The $Ni-Cr_2O_3$ catalyst is about 2·4 more active for the exchange reaction H_2-H_2O than for H_2-NH_3. For both reactions, chromia has a strong promoting effect, enhancing the activity per gram of catalyst of about three orders of magnitude in comparison with that of the Nickel black.

Die Aktivität eines $Ni-Cr_2O_3$-Katalysators für den Wasserstoffisotopenaustausch zwischen H_2-NH_3 und zwischen H_2-H_2O wurde in Abhängigkeit von der Reaktionstemperatur und den Partialdrücken von NH_3 und H_2O-Dampf gemessen. In beiden Fällen zeigte die Aktivität in Abhängigkeit vom Partialdruck ein Maximum; das Maximum im System H_2-NH_3 lag zwischen den Partialdrücken von 0,05 und 0,25 und im System H_2-H_2O zwischen 0,25 und 0,5. Der $Ni-Cr_2O_3$-Katalysator ist ungefähr um den Faktor 2,5 aktiver im System H_2-H_2O als im System H_2-NH_3. Für beide Reaktionssysteme zeigt Cr_2O_3 einen stark beschleunigten Einfluß, wobei die Aktivität pro Gramm des Katalysators um ungefähr 3 Größenordnungen im Vergleich zum Nickel-Schwarz gesteigert wird.

Keywords

activation energy; ammonia; catalysts; chemical reaction kinetics; chromiun oxide; hydrogen isotopes; isotopic exchange; nickel

1. Introduction

The isotopic exchange reaction of deuterium between hydrogen and ammonia

$$NH_3 + HD \rightleftarrows NH_2D + H_2 \quad (1)$$

in the absence of a catalyst takes place with a measurable rate only at temperatures higher than 600 °C [1]. In the presence of an appropriate catalyst, as the group eighth metals, that chemisorb dissociatively the hydrogen molecule, the reaction proceeds with high rate, even at room temperature [2, 3]. As catalysts for this reaction have been checked unsupported metals [4, 5], metals on supports [6, 7] and intermetallic compounds [8]. Depending on the chemical nature of the metal, the catalytic activity varies in a range of several orders of magnitude, the most active being platinum and the least active, iron. The catalytic activity for the last is about four orders of magnitude lower than for platinum [4]. In the literature, no information is available so far, referring to the influence of the support nature on the catalytic activity of the metals for the reaction (1).

The metals of the group eighth are catalytic active, at the same time, for the exchange reaction of deuterium between hydrogen and water [9–14],

$$H_2 + HDO \rightleftarrows HD + H_2O \quad (2)$$

As in the reaction (1), the most active metal for the exchange H_2-H_2O is platinum and the least active is iron [13]. With respect to nickel catalysts for the reaction (2) it has been evidenced a remarkable enhancement of the catalytic activity per gram of catalyst of about 2–3 orders of magnitude [15–20] when the metal is supported by chromia.

The question arises whether the promoting effect produced by chromia on the activity of nickel takes place as well in the case of the exchange reaction between hydrogen and ammonia.

The purpose of this paper is to evidence the promoting action of chromia on nickel activity for reaction (1). It is presented the results concerning the activity of $Ni-Cr_2O_3$ catalyst for the two isotopic exchange reactions in dependence on the temperature and on the ratio of the two reactants of the reaction mixture.

[1]) Address: R-3400 Cluj-Napoca 5, P.O. Box 700, Romania

2. Experimental

2.1. *Materials*

The $Ni-Cr_2O_3$ catalyst of composition Ni85 Cr15 at% was prepared by coprecipitation of a solution of chromium and nickel nitrate mixture with sodium carbonate solution. After filtration the precipitate was washed with twice destilled water, dried at 105 °C, calcinated in flowing nitrogen at 340 °C and reduced in flowing hydrogen at 350 °C. The detailed procedure has been described previously [20].

The deuterated hydrogen used for kinetic measurements for the reaction (1) was prepared by catalyzed isotopic exchange reaction between hydrogen and heavy water vapour and stored in a tank after careful drying. The deuterium content of the hydrogen was about 2 at% D. The ammonia for kinetic measurements have been p.a. and with natural content of deuterium, about 119·5 ppm.

For the experiments concerning reaction (2) it has been used electrolytic hydrogen containing about 50 ppm deuterium and a solution of heavy water of about 4 at% deuterium.

2.2. *Kinetic measurements*

The activity measurements of the catalyst for deuterium exchange reaction between hydrogen and ammonia (reaction 1) have been carried out at atmospheric pressure, in an installation represented in Fig. 1. It consisted of an all glass dynamic reactor containing the catalyst bed, introduced in a thermostatic bath. Before each kinetic measurement, the catalyst sample was reactivated "in situ" with hydrogen for 8 h at 150 °C. After that the reactant flows are measured by the two flowmeters, F_1 and F_2 and mixed in desired proportion; then the stream was carried through the catalyst bed. After reaction the gaseous mixture passed through a trap cooled with liquid nitrogen in order to condense ammonia. The deuterium content of the hydrogen stream has been determined by gas chromatography using a thermoconductivity detector.

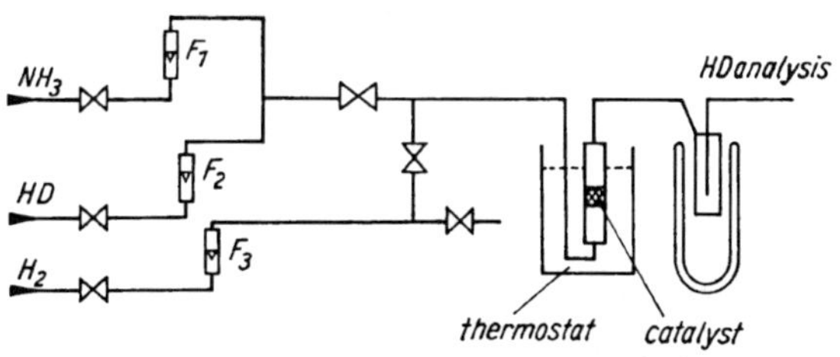

Fig. 1. Installation for kinetic measurements of deuterium exchange reaction between ammonia and hydrogen

Kinetic measurements for deuterium exchange reaction between hydrogen and water vapour (reaction 2) were carried out in a dynamic reactor as was described previously [21]. After the reactivation of the catalyst, the hydrogen flow, containing about 50 ppm deuterium, bubbled through water solution of about 4 at% D, then passed through the catalyst bed. The saturation and reaction temperatures were maintained constant with two thermostats. After reaction the hydrogen stream was dried and its deuterium content analysed by gas chromatography.

As *Varshavskii* and *Vaisberg* demonstrated, for small deuterium content, the exchange reaction (1) can be treated as a first order reaction [22]. In this case, the reaction rate r (expressed in moles of deuterium transferred per gram of catalyst during a second) is determined by the relation (3), as has been showed in [13]

$$r = \frac{G}{w(\alpha + \beta)} \ln \frac{n_0 - n_e}{n - n_e}, \tag{3}$$

where G is the flow rate of the hydrogen through the catalyst bed expressed in moles/s; w is the weight in grams of the catalyst sample; α is the value of the equilibrium constant [23] for the exchange reaction (1); β is the reduced partial pressure ratio of the reagents, ($\beta = \frac{2}{3} P_{H_2}/P_{NH_3}$); n_0 and n are the values of deuterium concentration of the hydrogen before and after reaction, respectively; n_e is the deuterium concentration in hydrogen at equilibrium. When the ammonia with natural isotopic content of deuterium is used, the equilibrium concentration n_e of hydrogen results from

$$n_e = \frac{\beta n_0}{\alpha + \beta}. \tag{4}$$

The isotopic concentrations n_0 and n have been obtained by direct measurements. The equilibrium value n_e has been calculated by means of the expression (4), but it was previously verified in two experiments and it was found that the difference between the measured and the calculated values didn't overpass 2%. The validity of the relation (3) has been also verified experimentally. The experimental conditions were so selected as to obtain a minimum value of the relative error for the reaction rate, r.

The catalytic activity in the case of the exchange reaction of deuterium between hydrogen and water vapour has been calculated by an expression similar to (3).

The measurements of the rate reaction have been performed in the kinetic region so that the reaction on the catalyst surface was the rate determining step, unaffected by the transport processes of the reagents to and from the catalyst surface. At the same time the experiments were carried out in those conditions where no capillary condensation of ammonia or water vapour could take place in the catalyst pores.

3. Results and discussion

3.1. *The influence of the temperature on the reaction rate*

The measurements were carried out in order to establish the reaction rate dependence on the temperature for the deuterium exchange between hydrogen and ammonia. The experiments have been performed for two values of the partial pressure of ammonia, χ, in the reactant mixture. The corresponding *Arrhenius* plots are illustrated in Fig. 2. The values of the activation energy have been evaluated as 11 kcal/mole for $\chi = 0{\cdot}085$ and 9·8 kcal/mole for $\chi = 0{\cdot}18$.

It should be pointed out that at reaction temperatures over 130 °C the experimental values of the reaction rate are deflected downwards from the straight line of the *Arrhenius* plot. This fact can be explained by the back diffusion process of the reaction products in the catalyst bed, occurring with increasing temperature on account of the high activity of the catalyst, as it has been mentioned elsewhere [21]. Therefore, all measurements were so conducted as to avoid the back diffusion disturbing effect on the reaction rate.

Mitteilungen / Information

IAEA: VERHANDLUNGEN UEBER KERNSICHERHEIT

Die erste Tagung einer Reihe von ausserordentlichen Tagungen ueber Kernsicherheitsfragen, die aufgrund des Reaktorunfalls in Tschernobyl abgehalten werden, wurde am 21.7.1986 im Amtssitz der IAEA in Wien eroeffnet. Etwa 150 Regierungssachverstaendige aus 56 Mitgliedsstaaten und neun internationalen Organisationen sind zusammengekommen, um die Entwuerfe zweier internationaler Abkommen zu besprechen. Durch das erste Abkommen wuerden die Vertragsparteien dazu verpflichtet, bei nuklearen Unfaellen mit moeglichen grenzueberschreitenden Auswirkungen Fruehwarnungen und Informationen zu geben; in diesem Abkommen wuerden die im vergangenen Jahr in der Informationsschrift INFCIRC/321 veroeffentlichten Richtlinien der IAEA beruecksichtigt. Durch das andere Abkommen wuerden die Vertragsparteien verpflichtet bei nuklearen Unfaellen alle Anstrengungen zu unternehmen, um Hilfestellung zu geben; in diesem Abkommen wuerden die 1984 in der Informationsschrift INFCIRC/310 veroeffentlichten Richtlinien beruecksichtigt. Falls ueber die Entwuerfe der Abkommenstexte Einigung erzielt wird, werden diese auf einer ausserordentlichen Tagung der IAEA-Generalkonferenz zur Annahme vorgelegt werden. Die ausserordentliche Tagung wird am 24.9.1986 im Kongresszentrum der Wiener Hofburg eroeffnet werden. Auf der Eroeffnungssitzung der Tagung stellte der Generaldirektor der IAEA, Dr. Hans Blix fest, dass fuehrende Politiker einer Reihe von Laendern, die die Kernenergie nutzen, zwar fuer die fortgesetzte und verstaerkte Nutzung der Kernenergie sind, sich aber sehr wohl fuer mehr internationale Zusammenarbeit zur Erhoehung der Kernsicherheit ausgesprochen haben. "In vielen Laendern wird das entschlossene Handeln der Regierung, alleine oder gemeinsam wie in der IAEA, dazu beitragen, die oeffentliche Meinung davon zu ueberzeugen, dass verschiedene Probleme, die erkannt wurden, zu bewaeltigen sind oder sogar schon bewaeltigt wurden", sagte der Generaldirektor. "Auf den Tagungen des IAEA-Gouverneursrates im Mai und Juni wurde eine Reihe von Massnahmen besprochen. Diese Tagung ist nur eine Massnahme unter vielen. Aber meiner Meinung nach ist es aeusserst wichtig, dass diese Tagung die ihr gesteckten Ziele erreicht".
Der am 21.7.1986 gewaehlte Vorsitzende der Vollversammlung der Tagung ist Seine Exzellenz Lodewijk H.J.B. van Gorkom, Botschafter in Oesterreich und Staendiger Vertreter der Niederlande bei der IAEA. Die Tagungen der Arbeitsgruppen ueber den Inhalt jedes der beiden Abkommen und ueber ihre Beziehung zueinander, werden von drei stellvertretenden Vorsitzenden geleitet werden: vom Botschafter Brasiliens, Seiner Exzellenz Carlos Augusto de Proenca Rosa, vom Botschafter Aegyptens, Seiner Exzellenz Mohamed El-Taher Shash, sowie vom stellvertretenden Staendigen Vertreter der DDR Herrn Joachim Maser. (IAEA-Pressekommunique (1986)16)

USA: NEW FOOD IRRADIATION RULE ADOPTED

A long-awaited rule stipulating requirements for low-dose radiation processing of foods has taken effect in the United States. As published in the US Federal Register by the Food and Drug Administration on 18 April 1986, the new rule requires all retail irradiated food products to carry a label clearly stating treatment with radiation and accompanied by the international symbol for the process.
At the wholesale level, irradiated food products also will require labels saying "treated with radiation - do not irradiate again" or similar wording. "The labelling requirement applies only to a food that has been irradiated, not to a food that merely contains an irradiated ingredient but that has not itself been irradiated," the rule says in part.
After two years, or as of 18 April 1988, the rule stipulates that all wording requirements for labelling irradiated food products will be dropped, leaving only the international symbol to inform consumers.
The new FDA rule allows fresh fruits and vegetables to be treated with radiation at up to 1 kilogray, and herbs and spices at up to 30 kilogray. Under previous rules, the FDA already allows radiation treatment of potatoes, wheat and wheat flour, food grade enzymes, and pork products for trichinae control. Previously, herbs and spices also could be treated with radiation for microbial and insect control at up to 1 kilogray. These levels are far below those that have been recommended as safe for all foods by the Codex Alimentarius Commission of the Food and Agriculture Organization and World Health Organization.
FDA Commissioner Frank E. Young called the new rule "a landmark development that will foster the use of this process, thereby providing better and safer products for the public and also enhancing the market potential for the producers, not only in this country but also for export". (IAEA-Bulletin 28(1986)2)

Kurznachrichten / Newlets

Neue analytische Methoden:

^{15}N-NMR-SPEKTROSKOPIE - NEUE METHODEN UND IHRE ANWENDUNG

Nach ^{1}H- und ^{13}C-Kernen ist in der Organischen und Bioorganischen Chemie der Stickstoff-Kern die wichtigste NMR-Sonde bei strukturchemischen Studien. Lange Zeit jedoch verzichtete man auf die Stickstoff-NMR-spektroskopische Untersuchung groesserer Molekuele - war sie doch mit einem zweifachen Handicap behaftet: Zum einen stoerte das elektrische Quadrupolmoment der ^{14}N-Kerne, zum anderen war die natuerliche Haeufigkeit des Isotops ^{15}N zu gering. Neue Aufnahmetechniken wie Pulssequenzen und Polarisationstransfer loesten zusammen mit der Verwendung groesserer Probenkoepfe und immer staerkerer Magnetfelder weitgehend das Problem der geringen NMR-Empfindlichkeit des ^{15}N-Kerns und bewirkten eine verhemente Entwicklung der ^{15}N-NMR-Spektroskopie. Diese gewaehrt inzwischen vielseitige Einblicke in die Molekuelstruktur in isotroper und anisotroper Phase. Ihre Anwendungspotential reicht von der Anorganischen ueber die Metallorganische und Organische Chemie bis hin zur Biochemie und Molekularbiologie. Untersucht werden z.B. reaktive Zwischenprodukte, der Metabolismus von Stickstoffverbindungen sowie Biopolymere und Enzym-Inhibitor-Komplexe, wobei gerade die Kombination mit 2D-NMR.Techniken besonders detaillierte Studien ermoeglicht. (Angew. Chemie, 98(1986)5, S.381-412, 184 Lit.-Stellen, W. von Philipsborn, R. Mueller)

Neue Methode zur Isotopentrennung:

MAGNETEFFEKTE BEI CHEMISCHEN REAKTIONEN

Lassen sich chemische Umwandlungen mit einem Magnetfeld steuern? Noch vor nicht allzu langer Zeit verneinten Fachleute diese Frage kategorisch. Zwar waren in der Fachliteratur von Zeit zu Zeit Hinweise auf den Einfluss von Magnetfeldern auf chemische Reaktionen zu finden, sie bestaetigten sich aber bei genauer Ueberpruefung nicht. Gut bekannt ist, dass die Energie der Wechselwirkung chemischer Teilchen mit Magnetfeldern sehr gering ist und gewoehnlich nur ein Zehntausendstel bis ein Millionstel der Waermeenergie der Teilchen betraegt. Daher war man auch der Ansicht, dass Magnetfelder chemische Reaktionen faktisch nicht beeinflussen koennen. Diese Anschauung, die sich fest eingebuergert hatte, die aber, wie sich herausstellte, falsch war, wurde durch die Arbeiten eines Kollektivs von Wissenschaftlern des Instituts fuer chemische Kinetik und Verbrennung der Sibirischen Abteilung der AdW der UdSSR und des Instituts fuer chemische Physik der AdW der

UdSSR widerlegt. Mit einer Serie von ueberzeugenden Experimenten und Berechnungen wies man nach, dass Magnetfelder chemische Reaktionen beeinflussen koennen und dies auch wirklich tun. Die Natur dieser Erscheinungen kann im Rahmen der klassischen Vorstellungen nicht erkannt werden, sie laesst sich aber mit der Quantentheorie der chemischen Reaktionen ganz natuerlich erklaeren. Es geht darum, dass der Quantenmechanik zufolge chemische Umwandlungen vom Elektronenspin der chemischen Teilchen abhaengen muessen. Und das Magnetfeld beeinflusst das Verhalten der Elektronenspins und ueber sie den Reaktionsablauf. Diese Entdeckung, der Einfluss des Magnetfeldes auf chemische Umwandlungen, gehoert zu den bedeutendsten Erkenntnissen der physikalischen Chemie in den vergangenen 25 Jahren. Sie erklaert mehrere bis dahin unbekannte Erscheinungen und erschliesst gleichzeitig neue Wege zur Steuerung einiger wichtiger chemischer und biochemischer Prozesse. Unter den voellig neuen Erscheinungen, die die Wissenschaftler entdeckten, sei besonders der sogenannte Magnetisotopieeffekt genannt. Auf der Basis dieses Effektes wurde eine neue Methode zur Isotopentrennung entwickelt. Ein entscheidender Beitrag zur Entwicklung der genannten neuen Richtung der modernen Chemie waren die Arbeiten des Akademiemitgliedes J.N. Molin und der Doktoren der Wissenschaften A.L. Butschatschenko, R.S. Sagdejew, K.M. Salichow und J.L. Frankewitsch. Fuer den Arbeitszyklus unter dem Titel "Magnet- und Spineffekte in chemischen Reaktionen wurde den Autoren der Leninpreis 1986 verliehen. (Presse der Sowjetunion (1986) 10)

Activation analysis:

THIN LAYER ACTIVATION FOR WEAR MEASUREMENT

Thin Layer Activation (TLA) is a highly sensitive method for the measurement of wear, erosion and corrosion. The technique, developed by Harwell is extensively used in research and development. Now, Cormon Ltd of Lancing, in collaboration with Harwell, is offering TLA to industry as a routine inspection and maintenance technique. Two feasibility studies for the Ministry of Defence on the application of TLA as a condition monitor of wear in gun barrels and on jet engine turbine blades were completed recently by Cormon. A third contract to prove the principle of using TLA to measure wear of high pressure turbine blades is currently being undertaken in collaboration with Rolls-Royce Ltd. TLA has been applied to a variety of components such as machine tools, cam followers, cylinder liners, valve inserts, piston rings and shaft bearings. Erosion has been measured in high pressure fuel pipes, fuel injectors and heat exchanger pipes. The activated area is small (about $3mm^2$ is typical) and the total activity exceedingly low. (Atom (1986) 357)

Kerntechnik:

CHINA: AUCH KAELTE AUS DER KERNENERGIE

In Guangdong und anderen warmen Provinzen des Landes will VR China die Kernenergie auch zur Kaelteproduktion nutzen, mit welcher Fernkuehlsysteme versorgt werden sollen. Dies geht aus einer kuerzlich veroeffentlichten Studie des Institute of Nuclear Energy Technology (INET) in Bejing hervor, der weiter zu entnehmen ist, dass fuer das Jahr 2000 der Einsatz von Heizreaktoren ausser in der Hauptstadt der VR China auch in den Städten Shenyang und Harbin geplant wird. Die nukleare Fernkuehlung, so fuehrt der Bericht praezisierend aus, ist nicht fuer private Haushalte, sondern fuer Laboratorien, Büros, Hotels und aehnliche Betriebe vorgesehen. Ein kleiner Prototyp-Reaktor mit 5 MW Leistung befindet sich im genannten Institut im Betrieb, waehrend in Suedchina ein Versuchsreaktor mit 7,5 MW Leistung speziell fuer die Verwendung zu Kuehlzwecken erprobt wird. In dieser Anlage laesst sich das zur Klimatisierung bestimmte Wasser auf eine Temperatur von 7 Grad C abkuehlen. (SVA-Bulletin (1986) 10/11)

Strahlentechnik:

USA: LEBENSMITTELBESTRAHLUNG MUSS SICH DURCHSETZEN

Nach der grundsaetzlichen Freigabe der Bestrahlung von frisch geerntetem Obst und Gemuese zur Bekaempfung des Schaedlingsbefalls und zur Verbesserung der Haltbarkeit hat die amerikanische Lebensmittelbehoerde, die Food and Drug Administration (FDA), nunmehr festgelegt, dass Frischprodukte mit einer Dosis bis zu 100 Kilorad und Trockenware wie Kraeuter und Gewuerze mit einer Dosis bis zu 300 Kilorad behandelt werden duerfen. Gemaess den neuen Bestimmungen muessen bestrahlte Lebensmittel durch den Vermerk "Treated with Radiation" gekennzeichnet sein. In der Auseinandersetzung ueber diese Bestimmung betonte die FDA, die Deklarationspflicht bedeutet keineswegs, dass Zweifel an der Unschaedlichkeit der Bestrahlung bestuenden. Seitens der Wirtschaft wurde indessen ausgefuehrt, die Konsumenten wuenschten wohl eher die zusaetzliche Information, dass die behandelten Produkte nicht radioaktiv seien. Wie rasch und wie umfassend sich die Strahlenbehandlung von Obst und Gemuese anwenden laesst, haengt auf jeden Fall entscheidend vom Verhalten der Oeffentlichkeit ab. Fachleute erwarten, dass sich die Lebensmittelbestrahlung auf dem Markt schrittweise durchsetzen wird. (SVA-Bulletin (1986) 10/11)

SULZER LIEFERT INDUSTRIELLE BESTRAHLUNGSANLAGE

Einen neuartigen Typ von Gamma-Bestrahlungsanlagen hat die Firma Beta-Gamma-Service (BGS), Wiehl BRD, bei Sulzer, Winterthur (Schweiz), bestellt. Die Anlage soll ausser zur Sterilisation von Industrieprodukten vorwiegend fuer die Modifikation von Kunststoff-Formteilen eingesetzt werden. Insbesondere durch Strahlenvernetzung werden wichtige Gebrauchseigenschaften von Polymeren verbessert. Die Bestrahlungsanlage, die fuer Aktivitaeten bis 3,5 MCi Co-60 ausgelegt ist, ermoeglicht unterschiedliche Dosen, wie sie fuer die Anwendung der Strahlenvernetzung erforderlich sind, gleichzeitiges Behandeln von Produkten, die mit unterschiedlichen Dosen bestrahlt werden muessen, sowie hohen Nutzungsgrad der Strahlenenergie. (SVA-Bulletin (1986) 10/11)

Umwelt:

MASSIVER RUECKGANG DER SCHWEFELDIOXID-EMISSIONEN IN FINNLAND

Die Emissionen von SO_2 aus der Energieproduktion in Finnland ging zwischen 1973 und 1984 um 40% zurueck. Dies geht aus der Praesentation von zwei Forschern an einem Seminar von Mitte Mai in Granz ueber die Kontrolle von SO_2 und NO_x hervor. H. Niininen und R. Penttila zeigten, dass im Gebiet des Energieverbrauchs in Finnland in der letzten Dekade eine rasche Umschichtung erfolgt ist, wobei der Anteil von Erdoel als Primaerenergiequelle von 55% auf 34% zurueckging. In der gleichen Periode stieg der Anteil von Kernkraft, Torf und Erdgas von praktisch 0% auf ueber 22% an. Der Anteil von Nuklearstrom an der gesamten finnischen Stromerzeugung betraegt zurzeit rund 40%. Der dramatische Rueckgang war im Sektor Elektrizitaet zu verzeichnen. Dort wurden die SO_2-Emissionen pro Produktionseinheit auf den fuenften Teil des Werts von 1973 abgesenkt. (SVA-Bulletin (1986) 10/11)

Veröffentlichungen/

Publications

CENTRAL RESEARCH INSTITUTE FOR PHYSIKS HUNG. ACADEMY SCI. BUDAPEST/HUNGARY

KFKI-1986-49K

Monitoring the radiation consequences due to the disaster at the Chernobyl nuclear facility from April 28 to June 12, 1986 - Preliminary report, in Englisch, A. Andrási et al

KFKI-1986-44/E

Submicron resolution amorphous halogenide optical grid - E. Hajtó; Gy. Zentai

KFKI-1986-38/F
Influence of temperature oscillation on measured crystal weight during Czochralski growth - By. Szabo; Z. Juhasz

ZENTRALINSTITUT FUER ISOTOPEN- UND STRAHLENFORSCHUNG DER ADW DER DDR, LEIPZIG

ZfI-Mitteilungen Nr.

107 Sorptionseigenschaften von Kationenaustauschern aus Loesungen hoher Elektrolytkonzentration - Ein Beitrag zum Problem der Selektivitaet von organischen Ionenaustauschern - G. Pfrepper, Nov.85

108 Rechentechnik und Informatik in der Kernforschung - KDT-Tagung, Roitzsch, 16.-18.10.84, Dez. 1985

109 Untersuchungen zur mikrobiellen Laugung von Zirkonen mittels Funkenquellen-Massenspektrometrie - S. Becker u.a.,

Programmsystem zur Behandlung hochangeregter Atomzustaende - M. Lorenz; E. Hartmann

Herstellung von Radiotracern fuer Anwendungen in der Technik - R. Otto, Dezember 1985

110 Methoden der MN-15-Tracerforschung in biologischen Systemen - K. Hirschberg H. Faust

^{15}N-^{n}X-Mehrfachmarkierungen bei Stoffwechseluntersuchungen am Menschen - P. Krumbiegel, Dezember 1985

111 Die permeative Abtrennung der Spaltedelgase Krypton und Xenon aus der Abluft kerntechnischer Anlagen - Chr. Jonas, Juni 1986

112 Wissenschaftlicher Ergebnisbericht 1985 - Juni 1986

113 Zeitaufloesende ESR-Spektroskopie - Grundlagen und Anwendungen in der Strahlen- und Photochemie - D. Beckert, Mai 1986

114 N-15 in Biological Nitrogen Fixation Studies - A Bibliography - H. Faust, Mai 1985

115 6. Diskussionstagung Massenspektrometrie Stolberg/Harz, 25.-29.11.85 - S. Becker; H.-J. Dietze, August 1986

116 Mass attenuation coefficient tables for X-rays - G. Brunner

Beitraege zum Zustand von Rh III in salzsaurer Loesung - L. Voigt; G. Pfrepper

Untersuchungen zur Abtrennung von Palladium und Caesium aus salpetersauren Loesungen und Abfalloesungen der Wiederaufbereitung von Kernbrennstoff durch Metallferro- und Metallferricyanide - R. Pfrepper u.a.

Moeglichkeiten der Abtrennung von Palladium und Silber aus salpetersauren Laugeloesungen von Elektronikschrott durch Faellungsverfahren - G. Pfrepper u.a.

Einsatz der Maximum-Likelihood Methode bei der Parameterschaetzung in radiometrischen Modellgleichungen - M. Braune

Variation of heavy-ion-induced X-ray emission by the chemical state of the target - E. Hartmann

Clusterionenbildung hexagonalen Bornitrids im Laserplasma-S.Becker;H.J.Dietze

Zur numerischen Bestimmung der Kernbrennstoffzusammensetzung - K. Zieger August 1986

117 Ausgewaehlte Strahlungsionisationsdetektoren fuer die Gaschromatographie: Grundlagen, Anwendungen und Perspektiven - P. Popp

Determination of inorganic gases using radiation ionization detectors - P. Popp; G. Oppermann, September 1986

UNITED KINGDOM ATOMIC ENERGY AUTHORITY

AERE-R 11974
Proceedings of the specialicts' workshop on iodine chemistry in reactor safety. September 1985, 365pp, A.M. Deane; P.E. Potter

AERE-M 3496
A pictorial view of radionuclide migration from a deep underground repository for cemented intermediate-level waste. November 1985, 22pp, A.W. Herbert et al

AERE-R 11865 RKFB
A reactor kinetic code with feedback. November 1985, 46pp, F. Lemay; J. Needham

AERE-R 11946
Improvements to the Harwell Van de Graaf accelerator. January 1986, 29pp, P.M. Read; J.H. Stevens

AERE-R 11998
A theoretical study of the uses of eddy current impedance methods for the measurement of claddings. February 1986, 31pp, R.P. Holt; K.D. Boness

AERE-R 12012
Fundamental physics research at Harwell in the 1950s. II. The Van de Graaf Group. March 1986, 31pp, J.M. Freemann

AERE-R 12023
A literature survey of the possible effects of humic and fulvic acids on the disposal of LLW and ILW. March 1986, 29pp, F.T. Eward; S.J. Williams

AERE-R 12133
The measurement of nitrogen dioxide in the outdoor environment using passive diffusion tube samplers. February 1986, 16pp, D.H.F. Atkins et al

AERE-R 12154
The mixing of aerosols during in-situ filter testing. March 1986, 16pp, J.L. Cunningham; B.L. Green

P R O C E E D I N G S

CENTRAL INSTITUTE OF ISOTOPES AND RADIATION RESEARCH OF THE ACADEMY OF SCIENCES OF THE GDR, LEIPZIG

Proceedings of the Third Working Meeting on "Radioisotope Application and Radiation Processing in Industry" held in Leipzig, 23-27 September, 1985, Ed. by D. Luther, 2 Volumes, 1458 pp, 137 lectures and contributions

CENTRAL INSTITUTE OF PHYSICS, BUCHAREST, ROMANIA

Proceedings of the XV Course of the Brasov International School in Physics, Poiana Brasov, Romania, August 28 - September 8, 1984

"First Part, Atomic Physics;, edited by Al. Berinde; I.A. Dorobantu and C. Zoran, 1986, 488 pp, 23 lectures and contributions

"Second Part Nuclear Physics", edited by G. Semenescu; I.A. Dorobantu and N.V. Zamfir, 1986, 769 pp, 32 lectures and contributions

Ausgewählte Beiträge / Recently published papers

ATOMNAJA ENERGIJA 61(1986)4 (all in Russian)
Investigation of the capture cross-section of structural elements by measuring the neutron balance of multiplying media - V.I. Golubev et al

Grain-boundary deformation in Ni irradiated with neutrons - Sh.Sh. Ibragimov

On designing counter-current separation cascades with arbitrary stage enrichment - E.B. Gruzdev et al

Contemporary tendencies of development in hydrometallurgical processes of uranium extraction from ores - D.I. Skorovarov

Absorption of gamma radiation energy in air - A.V. Zhemerev

Background of a liquid-xenon scintillation spectrometer - I.R. Barabanov et al

Calculation of spatial energy-angular field characteristic for a point isotopic gamma-sources in an infinite air medium - V.S. Gorev et al

Composition thermoluminescent systems for measuring photon radiation energy - L.Z. Kalmykov; G.I. Gorelik

Neutron activation analysis of Au and accompanying elements in geological symples using a power reactor - Yu. N. Burmistenko et al

Obtaining the short-life isotope I-123 on Microtron MT-22 - G.N. Flyorov

Cesium deposition on surfaces of a sodium coolant circuit - V.D. Kizin et al

ANGEW. CHEMIE 98(1986)4
Adolf von Baeyers wissenschaftliches Werk - ein Vermaechtnis - R. Huisgen

Das Konzept der Spannung in der organischen Chemie - K.B. Wiberg

Geordnete Bereiche in verduennten Loesungen von ionischen Spezies durch Anziehung gleichsinnig geladener Teilchen - ein Paradigmenwechsel in der Kolloid- und Polymerchemie - N. Ise

Von Triphenylamin abgeleitete bicyclische Wirthohlraeume - Gastselektivitaet und Rodoxeigenschaften - H. Schrage et al

REPROCESSING NEWS (1986) 8
Stand der Entsorgung verbrauchter Brennstoffe in (West) Europa

Termingerechter Ausbau der Wiederaufbereitungsanlage in La Hague (Frankreich)

Vollstaendige Schliessung des Brennstoff-Kreislaufs durch U-Rezyklierung

Stoerfaelle in Sellafield

LA RECHERCHE (1986)173 (in French)
EXAFS: How to see the environment of atoms - P. Lagarde

LA RECHERCHE (1986)174 (in French)
The formation of the universe - Trink Xuan Thuan

The pseudofiction of nuclear apocalypse - J. Vilna

LA RECHERCHE (1986)175 (in French)
Substitute fuels - G. de Gaudemaris et al

Kekule's snake - from dream to science - J.-M. Levy - Leblond

Some hard-to-fool neutrinos - F. Vannucci

LA RECHERCHE (1986)176 (in French)
Inflation of the universe - A. Bouquet

The origin of agriculture in Europa - R.W. Dennell

The triton: a nuclear mirror - J.M. Cavedon et al

LA RECHERCHE (1986)177 (in French)
Protecting the genetic heritage - M.Hofnung

The messengers of immunity - D. Fradelizi

Antibody diversity - F. Rougern

Modern vaccines - A. Sasson

LA RECHERCHE (1986)178 (in French)
The Antarctic Ocean - P. Treguer; G. Jacques

Quasi - crystals - G. Gratias

Wave and particle: The double face of quantum physics - D. Tarnowski

Nuclear energy in Sovjet Union - G.Matcheret

ECONOMIC GEOLOGY 81(1986)3
High O-18 ore-forming fluids in volcanic-hosted base metal massive sulfide deposits: Geologic, O-18/O-16, and D/H evidence from the Iberian belt, Crandon, Wisconsin, and Blue Hill, Main - J. Munha et al

Pb and Sr isotope composition of hydrothermal minerals from the Great Bear Lake silver deposits, NWT, Canada - A. Changkakoti et al

Z. NATURFORSCHUNG 41A(1986)5
Cu isotope separation by ion exchange electromigration and Cu isotope analysis by thermo-ionization mass spectrometry - Y. Fujii et al

ANALYTIC. CHEM. 58(1986)7
Inductively coupled plasma mass spectrometry applied to isotopic analysis of Fe in human fecal matter - B.T.G. Ting; M.Janghorbani

Isotope - selective trace - element detection with the thermoionic diode - K. Niemax et al

GEOKHIMIYA (1986) 5 (in Russian)
Sm-Nd isotopic dating of the most ancient rocks of the Ukrainian shield and Omolon massif - E.V. Bibikova, H. Baadsgaard

Isotopic composition of Th in volcanic rocks of contionental rift zones - A.J. Polyakov et al

RADIOCHIMICA ACTA 39(1986)1
Dissolution of Np dioxides in aqueous solutions under various conditions - K.H. Lieser et al

Accuracy and precision in NAA of U - Louis F. Beltido

Thiosemicarbazone complexes of Tc - F. Graser; C. Genestar

CHEM. PHYSICS LETT. 127(1986)4
A surface hopping quasiclassical trajectory study of the $H_2^+ + H_2$ and $(H_2 + D_2)^+$ systems - C.W. Eaker; G.S. Schatz

Bimolecular radical scavenging processes for laser isotope separation : $NH_2D + O_2$ - C.T. Lin

DOKLADI AKADEMII NAUK 288(1986)3 (in Russian)
Thermodiffusion speration of gas mixtures in a sound wave field - V.B. Bozhdankevich et al

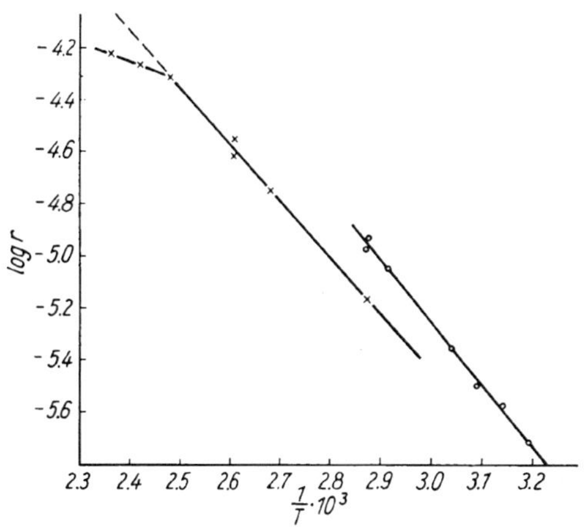

Fig. 2. *Arrhenius* plots for NH_3-H_2 deuterium exchange reaction (○ corresponding to $\chi = 0{\cdot}085$; × corresponding to $\chi = 0{\cdot}18$)

For the exchange reaction of deuterium between hydrogen and water vapour, as it has been shown elsewhere [24] the activation energy values extended on a wide interval subjected to a compensation effect. However, in most cases, the activation energy values were include in the interval of 13 to 15 kcal/mole, that are significantly higher than those obtained for the reaction (1).

3.2. The influence of the reactant ratio on the reaction rate

For the deuterium exchange reaction between hydrogen and ammonia the reaction rate measurements were performed in dependence on the partial pressure of ammonia χ. The partial pressure has been varied from 0·036 to 0·79, while the reaction temperature was kept constant at 50 °C. The results are represented in Fig. 3. It can be seen that the reaction rate depends strongly on the ammonia partial pressure, presenting a maximum for values of χ between 0·05 and 0·25. After that r value decreases as the ammonia partial pressure is increased.

A similar behaviour of the rate of the reaction (1) has been observed for platinum [6] and iron [25] catalysts. In both cases the results have been interpreted as a proof of a *Langmuir-Hinshelwood* mechanism, the isotopic exchange reaction occurring between the two reactants adsorbed on the catalyst surface. The assumption is made that the reaction rate is proportional to the product of the covering degrees on the catalyst surface.

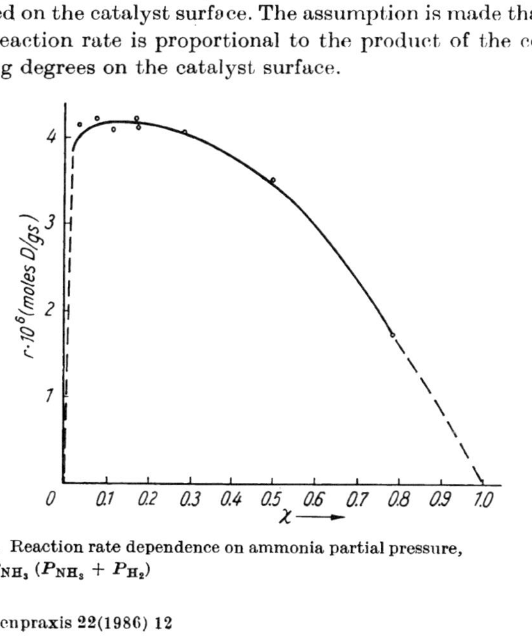

Fig. 3. Reaction rate dependence on ammonia partial pressure, $\chi = P_{NH_3}(P_{NH_3} + P_{H_2})$

For the deuterium exchange reaction between hydrogen and water vapour (reaction 2) catalyzed by $Ni-Cr_2O_3$ it has been shown in literature [26] that the dependence of the catalytic activity on the reactant ratio follows also a curve with a maximum as given in Fig. 4. The measurements were performed at the reaction temperature of 109 °C in order to avoid capillary condensation of the water vapour in the catalyst pores. As it can be seen, the maximum of the activity is placed for χ values between 0·25 and 0·5. Similarly, it may be assumed that the reaction (2) follows a *Langmuir-Hinshelwood* mechanism, too.

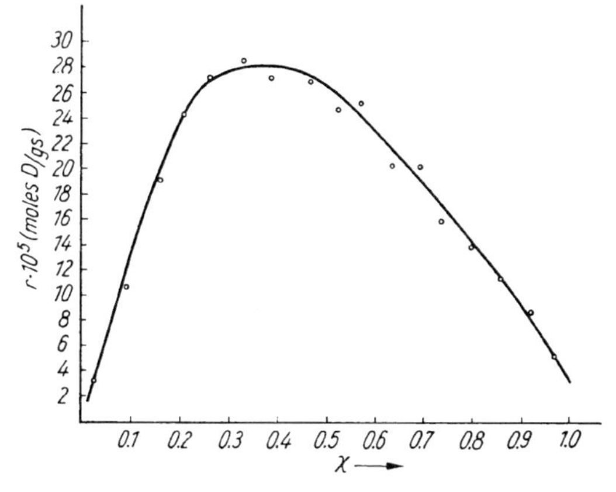

Fig. 4. Reaction rate dependence on water vapour partial pressure, $\chi = P_{H_2O}/(P_{H_2O} + P_{H_2})$, [26]

For the deuterium exchange reaction between hydrogen and water vapour (reaction 2) it has been, also, proposed another possible mechanism by *Temkin* and coworkers [27, 28]. They have assumed that the reaction (2) occurs according to a two-step mechanism by i) the oxidation of the nickel surface with water followed by ii) the reduction of nickel oxide with hydrogen. As it has been shown in [24] the experimental data of r in dependence on the partial pressure of water vapour verify the *Temkin*'s mechanism in a wide range of χ values.

In the case of the deuterium exchange between hydrogen and ammonia the data given in Fig. 3 have been transformed in the coordinates of the *Temkin*'s mechanism. They verified the validity of this mechanism in a wide range of χ values, too.

From kinetic data presented above it would result that the two mechanisms could be valid for both exchange reactions. Therefore, the available kinetic results are not conclusive to support one, or the other, of the mechanisms suggested.

3.3. Comparison between the catalytic activities of the two exchange reactions

In order to compare the values of the reaction rate for the two exchange reactions, parallel kinetic measurements have been undertaken on identical samples of $Ni-Cr_2O_3$ catalyst in similar experimental conditions of temperature and ratio of reactants.

In Tab. 1 are illustrated the working conditions and the values of the reaction rate on Nickel-Chromia catalyst. The last column presents the ratio of the two rates, r_{H_2O}/r_{NH_3}. The data show that $Ni-Cr_2O_3$ catalyst has a higher activity for the isotopic exchange reaction between hydrogen and water vapour than for the reaction between hydrogen and ammonia. The ratio of the two rates is laid between 2 and 3 with an average value of 2·4.

Tab. 1. Data concerning the activity of the $Ni-Cr_2O_3$ catalyst in the H_2-NH_3 and H_2-H_2O deuterium exchange reactions

H_2-NH_3			H_2-H_2O			Reaction rate ratio r_{H_2O}/r_{NH_3}
NH_3 partial pressure χ	Reaction temperature °C	Reaction rate $r.10^5$ moles D/g. s	H_2O vapour partial pressure χ	Reaction temperature °C	Reaction rate $r.10^5$ moles D/g. s	
0·036	50	0·42	0·032	50	0·83	1·98
0·085	65	0·84	0·085	65	1·85	2·20
0·085	70	0·90	0·085	70	2·60	2·89
0·085	75	1·18	0·085	75	3·10	2·63
0·18	100	1·99	0·17	100	4·13	2·08
					average value	2·4

In the case of the exchange reaction between hydrogen and water vapour it has been shown in the literature [15 to 20] that chromia has a strong promoting effect on nickel activity enhancing the r value about 2–3 orders of magnitude.

Our experiments showed neither the nickel oxide and chromia or chromia-nickel oxide mixture, have any catalytic activity for the two isotopic exchange reactions under given conditions as described above.

In order to verify whether chromia has a promoting effect on the nickel activity in the case of the H_2-NH_3 deuterium exchange too, it has been checked the catalytic activity of the nickel black sample for this reaction. The nickel black catalyst has been prepared by precipitation of nickel nitrate solution with sodium carbonate. Further, the procedure followed the method described above for the nickel-chromia catalyst preparation. The catalytic activity of the nickel black was measured in the conditions $\chi = 0{\cdot}18$ and the reaction temperature 75 °C; it was obtained for the value of the reaction rate $1{\cdot}3 \times 10^{-8}$ moles D/g·s.

From Tab. 1 it can be seen that at 75 °C and for $\chi = 0{\cdot}085$ the catalytic activity of $Ni-Cr_2O_3$ for the reaction (1) is $1{\cdot}18 \times 10^{-5}$ moles D/g·s, that means with about three orders of magnitude higher than the catalytic activity of nickel black. These data emphasize that Cr_2O_3 has also a strong promoting effect on the catalytic activity of the nickel for the H_2-NH_3 deuterium exchange reaction.

Both the shape of the curve representing the reaction rate versus partial pressure of ammonia or water vapour and the support effect given by chromia on the catalytic activity of nickel lead to the conclusion that the two reactions are ruled by a similar mechanism. On the basis of the kinetic data discussed above it is difficult to decide which of the two mechanisms is the most likely. As it has been mentioned in [29, 30] the participation of the support surface to the activation of water molecule for the isotopic exchange reaction between hydrogen and water is possible. Similarly, the chromia could activate the ammonia molecule for reaction (1). Thus, at the metal-support frontier are created conditions in which the reaction rate is much enhanced, in comparison with that of nickel black, by appearance of new catalytic sites much more active. This fact leads to a strong promoting effect of the chromia on the catalytic activity of nickel for both isotopic exchange reactions.

Acknowledgements

The authors would like to thank Mr. *Nicolae Peteanu* for experimental help, and Dr. *Romulus V. Bucur* for providing very helpful comments. This work was sponsored by the State Commitee for Nuclear Energy.

References

[1] *A. Farkas*, J. Chem. Soc. (1936) 26
[2] *G. C. Bond*, "Catalysis by Metals", Academic Press. London/New York 1962, p. 223
[3] *A. Ozaki*, "Isotopic studies of heterogeneous catalysis", Kodansha Ltd., Academic Press: Tokyo/New York/San Francisco/London 1977, p. 53
[4] *C. Kemball*, Trans. Faraday Soc. **48** (1952) 254; Proc. Roy. Soc. A **214** (1954) 413
[5] *Mijahara Koshiro*, Ref. Zhur. Khimia (1963) Nr. 3, B 486; (1965) Nr. 8, B 767
[6] *R. Haul, D. Blennemann*, J. Catal. **1** (1962) 432
[7] *U. Hiroshi, T. Naoyuki*, Bull. Chem. Soc. Japan **29** (1956) 762
[8] *G. Ya. Ryskin, Yu. M. Vaikov* et al., Ref. Zhur. Khimia (1982) Nr. 20, B 1143
[9] *K. Hirota, J. Horiuti*, Sci. Pap. Inst. Phys. Chem. Res. (Japan) **30** (1936) 151
[10] *N. H. Sagert, R. M. L. Pouteau*, Canad. J. Chem. **49** (1971) 3411; **50** (1972) 3686; **51** (1973) 4031; **52** (1974) 2960; Platinum Met. Rev. **19** (1975) 16
[11] *W. G. McNaught, C. Kemball, H. F. Leach*, J. Catal. **34** (1974) 98
[12] *G. G. Joris*, Catalytic Exchange: Platinum and Palladium Catalysts, in: "Production of Heavy Water" (*M. G. Murphy, C. H. Urey, I. Kirschenbaum*, eds.). Mc Graw-Hill: New York 1955, p. 222
[13] *B. Brigoli, E. Cerrai, R. Renzoni, M. Silvestri*, Nuovo Cimento Suppl. **5** (1957) 479
[14] *I. Ursu, D. Strugaru, O. Piringer*, Studii Cercet. Fiz. (Bucharest) **14** (1963) 421
[15] *E. H. Taylor*, Development of Nickel-Chromium Oxide as a Catalyst for Isotopic Exchange between Hydrogen and Water, in: "Production of Heavy Water" (*M. G. Murphy, C. H. Urey, I. Kirschenbaum*, eds.). McGraw-Hill: New York 1955, p. 154
[16] *D. Barb*, Studii Cercet. Fiz. (Bucharest) **17** (1965) 383
[17] *I. Ursu, D. Barb, M. Bologa, I. Hodor*, Unpublished data.
[18] *A. Olariu*, Studii Cercet. Fiz. (Bucharest) **22** (1970) 953
[19] *N. H. Sagert, P. E. Show-Wood, R. M. L. Pouteau*, Canad. J. Chem. **53** (1975) 3257
[20] *P. Mărginean, A. Olariu*, J. Catal. **8** (1967) 359
[21] *N. Palibroda, A. Olariu, P. Mărginean*, J. Catal. **12** (1968) 347
[22] *Ya. M. Varshavskii, S. Z. Vaisberg*, Zhur. Fiz. Khimii **29** (1955) 523; **32** (1958) 454
[23] *J. Bigeleisen*, The Significance of the Product and Sum Rules to Isotope Fractionation Processes, in: "Proceedings of the International Symposium on Isotope Separation, Amsterdam 1957" (*J. Kistenmaker, J. Bigeleisen, A. O. C. Nier*, eds.). North-Holland: Amsterdam 1958, p. 121
[24] *N. Marazan, P. Mărginean, A. Olariu*, Abhandlung der deutsch. Akad. der Wissenschaften Berlin, Kl. Geol. Biol. (1964) Nr. 7, 811
[25] *J. Weber, K. J. Laidler*, J. Chem. Phys. **19** (1951) 1089
[26] *T. Fodor, P. Mărginean, A. Olariu*, Studii. Cercet. Fiz. (Bucharest) **22** (1970) 697
[27] *M. I. Temkin*, Zhur. Fiz. Khimii **31** (1957) 3
[28] *M. L. Nachmanovich, N. M. Morozov, L. G. Buadze, M. I. Temkin*, Dokl. Akad. Bauk SSSR **148** (1963) 1346
[29] *J. H. Rolston, J. W. Goodale*, Canad. J. Chem. **50** (1972) 1900
[30] *P. Mărginean, A. Olariu*, J. Catal. **95** (1985) 1

Isotopenpraxis 22 (1986) 12, S. 439 bis 443

Environmental Application of Radiation Processing*)

A. K. Pikaev

(Academy of Sciences of the USSR, Institute of Physical Chemistry, Moscow)[1])

The paper is the review of current state of the application of radiation chemistry and radiation technology for environmental conservation. Some aspects of radiation treatment of natural water, radiation purification of aqueous wastes, radiation treatment of sewage sludge and radiation purification of exhaust gases are considered. The special attention is paid to the works of Soviet authors.

Die Arbeit behandelt den gegenwärtigen Stand der Anwendung von radiochemischen und strahlentechnischen Methoden auf dem Gebiet des Umweltschutzes. Es werden einige Gesichtspunkte der Strahlenbehandlung natürlicher Wässer, der Reinigung von wäßrigen Abfällen und Abfallschlämmen mittels Strahlung sowie der Reinigung von Abgasen betrachtet. Dabei stehen Arbeiten sowjetischer Autoren im Vordergrund.

Keywords

environmental engineering; flue gas; gaseous wastes; irradiation procedures; liquid wastes; radiation chemistry; sewage sludge; waste processing

Introduction

It is well-known that during last decades the problem of environmental conservation was sharpened. Because of destructing and sterilizing action of ionizing radiation the radiation methods can play an important role in the solution of this problem. It caused the increasing attention to using the radiation chemistry and the radiation technology in this field. At present numerous literature including books [1–5], booklets [6, 7], many reviews (see e.g. [8–12]) and special issue of "Radiation physics and chemistry,, [13] is devoted to the application of radiation methods for environmental conservation.

Investigations and technological developments in the field under consideration are carried out in the following main directions:

1) radiation treatment of natural water;
2) radiation purification of industrial wastes;
3) radiation treatment of sewage sludge;
4) radiation purification of exhaust gases.

The works in the first three directions are accompanied by the studies of disinfection of irradiating objects. In some cases (for example, in the case of wastes of live-stock breeding farms) the radiation treatment is used mainly for the disinfection.

Radiation treatment of natural water

Natural water before applications is purified mainly from organic compounds causing colour, odour etc. Usually the water is purified by the following way. At first it is treated by coagulant, then by lime (in the periods of insufficient alkalinity).

A suspension formed after the coagulation is settled and filtered off. The disinfection of water is performed by chlorine, chlorinated lime or ozone. As it is seen, this technological scheme is comparatively complex. Besides, the removal of organic compounds is incomplete [9]. It was a main reason for the proposition (see [1]) to use ionizing radiation for the decomposition of these organic compounds. The studies carried out in this direction have shown [1, 9, 11] that even at comparatively low doses (ca. 1 kGy) decolouration, deodorization and disinfection of water up to level corresponding to U.S.S.R. State Standard "Drinking Water" have taken place.

The data on the decolouration of water from reservoirs of Moscow District via the treatment by ^{60}Co gamma-rays are listed as an example in Tab. 1. The colour of water is measured in degrees; for drinking water it should be ≦ 20 degrees. From the table it follows that the irradiation is effective method for colour removal; radiation effect is increased by bubbling the air through the water.

The colour removal by radiation treatment is caused mainly by the decomposition of humus compounds which are responsible for the colour. Their decomposition occurs under the action of water radiolysis products (mainly OH radicals).

It was obtained also (see [9]) that sufficient deodorization was achieved at dose 1 kGy. At this dose the considerable decrease of virus intoxication and helminthization of the

Tab. 1. Radiation decolouration of water from reservoirs of Moscow District*

Reservoir	Initial colour, deg.	D_{20}, kGy	
		Air bubbling	Without air bubbling
Ucha reservoir	35	0·6	0·8
Eastern water treatment plant	33	0·35	0·5
Klyaz'ma reservoir	49	0·8	4

* D_{20} is the dose required for colour reduction to 20 degrees

*) Presented at the 3rd Working Meeting on "Radioisotope Application and Radiation Processing in Industry" held in Leipzig, 23–27 September 1985

[1]) Address: SU-117915, Moscow, Leninsky Prospekt 31, U.S.S.R.

water occurred [9]. Therefore the irradiation of natural water at dose 1 kGy provides its complex purification.

On the basis of the results considered the without-reagent scheme for purification and disinfection of natural water was proposed [9]. The scheme consists of two stages. The first stage is the removal of solid particles by means of microfilters; the second one is the irradiation of the water.

Hence the comparatively low doses are required for the purification of natural water. Because of it the radiation treatment can be considered as promising method (especially in the case of application of high-power accelerators). The simple calculation shows that the purification of natural water, for example, for water-supply of the city with 100 thousands inhabitants at water consumption of 10 l a hour for 1 man can be performed by means of 1 electron accelerator with power of $\approx 0{\cdot}5$ MW or by 5 accelerators with power ≈ 100 kW each (at dose 1 kGy and coefficient of using the radiation 0·6).

Radiation purification of industrial liquid wastes

The composition of industrial wastes depends on the type of the production. Often the wastes contain the compounds which are poisonous or hardly decomposed (e.g. cyanides, salts of mercury and bismuth, phenol etc.). Their concentrations can be comparatively high, and the doses for their decomposition can be relatively high ($\geqq 10$ kGy). As an example let us discuss the situation with wastes containing phenol.

According to [14], at gamma radiolysis of 10^{-3} M solution of phenol $G(-C_6H_5OH)$ are 2·66 and 0·45 molec per 100 eV in the presence and the absence of air, respectively. Therefore the dose required for the decomposition of 10^{-3} mole/l of phenol is 3·62 kGy. However, phenol is transformed into hydroquinone, pyrocatehol, hydroxyquinone and resorcinol, which should be decomposed also. Final products of the multi-stage transformation are maleic acid and other non-aromatic compounds. The dose for such transformation is high (0·2–0·25 MGy) (see [11]).

Apparently, such high doses is the considerable hindrance for industrial application of radiation treatment of wastes. Because of it recently the combined methods began to develop for the reduction of required doses. In such methods the radiation treatment is conjucted with conventional method (chemical, biological, flotational etc.) [11, 14–16]. Let us consider shortly these combined methods.

Radiation tertiary treatment. It was shown by several authors (see e.g. [11, 16–18]) that in some cases the radiation tertiary treatment of wastes after their chemical and biological purification occurred at low doses (tenth fractions of kilogray). For example, it was found [11] that the doses 0·1–0·3 kGy (in conjunction with air bubbling) were required for tertiary treatment of wastes from Baikal cellulose-paper plant.

Radiation-flotation purification. An example of this method is the radiation-flotation purification of industrial wastes from mercury developed in work [18]. The wastes were from the plants producing chlorine and alkali by electrolysis with using the mercury as an electrode. The method consists of three stages:

1) preliminary flotation in the presence of surfactant (sodium alkylsulfonate);
2) gamma irradiation;
3) secondary flotation (also in the presence of sodium alkylsulfonate).

The principal scheme of the method is shown in Fig. 1. By this combined method it is possible to purify the wastes from mercury presenting in different forms (dissolved, as a metal and in precipitate) up to limiting permitted concentration ($5 \cdot 10^{-3}$ mg/l). Some respective data are shown in Fig. 2.

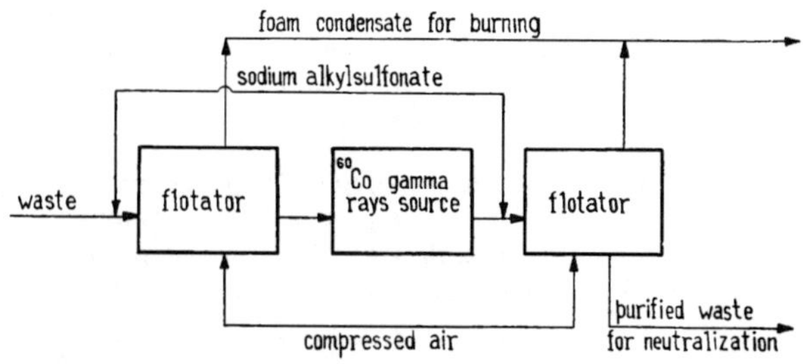

Fig. 1. Principal scheme of purification of wastes from mercury by radiation-flotation method

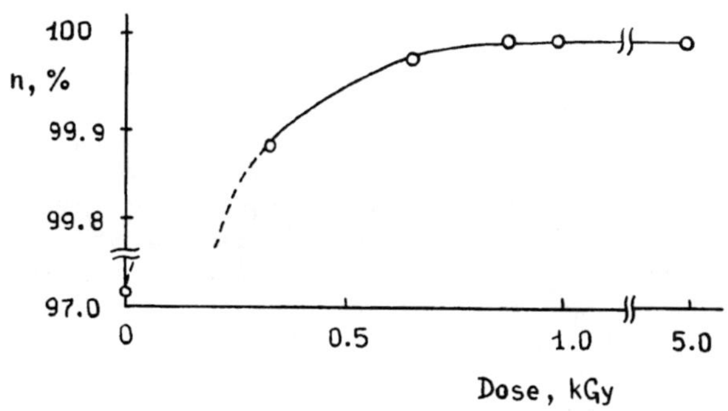

Fig. 2. The dependence of purification degree (n) on dose (initial content of mercury: metallic – 0·18 mg/l, ionic – 0·06 mg/l, in precipitate – 9·16 mg/l)

Optimum conditions for the purification are the following:

1) at preliminary flotation the pH value of waste is 12 to 13; concentration of sodium alkylsulfonate is 50 mg/l; the rate of air feed is 0·02–6·0 l/min;
2) at gamma irradiation the pH value is also 12–13; dose is 1 kGy;
3) the conditions for secondary flotation are the same as for preliminary one.

The mechanism of the influence of ionizing radiation on the purification process is the following. Obviously, the reduction of dissolved bivalent mercury which is not flotated to metallic one and/or weakly soluble pseudomonovalent form ($Hg_2(OH)_2$ type) which are removed effectively by flotation occurs during gamma irradiation. At dose 1 kGy the oxygen dissolved in waste is consumed completely, and in alkaline medium "ionic" mercury is reduced by hydrated electrons e_{aq}^-:

$$H_2O \rightsquigarrow e_{aq}^-, H, OH, H_2, H_2O_2\,, \quad (1)$$

$$H + OH^- \rightarrow e_{aq}^- + H_2O\,, \quad (2)$$

$$Hg(II) + e_{aq}^- \rightarrow Hg(I)\,. \quad (3)$$

It is confirmed by the data obtained from the study of microsecond pulse radiolysis of alkaline solutions of "ionic" and metallic mercury [19, 20].

It was found that under the action of the electron pulse short-lived optical absorption in UV-region appeared in such solutions. As an example this absorption in deaerated $\approx 10^{-4}$ M solution of HgO in 0·1 M KOH is shown in Fig. 3 [20]. The absorption belongs to Hg(I). In the presence of metallic mercury Hg(I) decays in pseudo-first order process:

$$Hg(I) + Hg(O) \rightarrow Hg(O) \cdot Hg(I) . \quad (4)$$

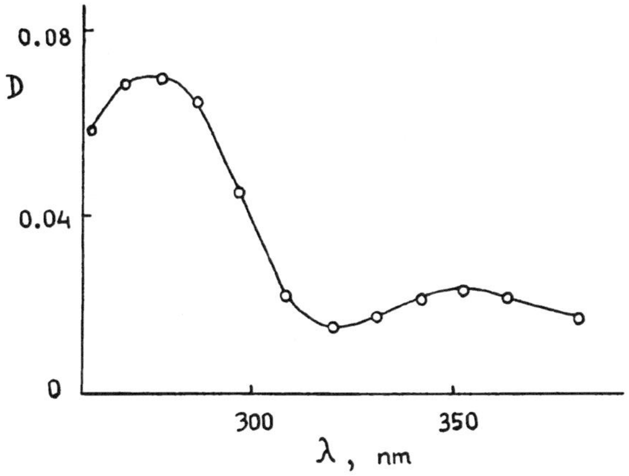

Fig. 3. Short-lived optical absorption in deaerated $\approx 10^{-4}$ M solution of HgO in 0·1 M KOH after microsecond electron pulse

The species Hg(O) · Hg(I) formed react then with each other giving rise to aggregated particles containing the products of dismutation and dimerization of Hg(I).

Radiation-adsorption purification. This method consists of the adsorption of pollutants from waste by active carbon during continuous irradiation of the sorbent (see [9, 11]). According to [9], the radiation-adsorption method is most effective for tertiary treatment (in particular, after biological purification).

Radiation-polymerization purification. The method is based on the capture of pollutants by insoluble precipitate of polymer formed during the irradiation of monomer solutions [2]. It seems to be perspective for the purification of wastes of polymer industry in which the addition of monomers is not necessary. For example, the data on successful purification (in laboratory experiments) of wastes of plants producing polyethers and polyvinyl chloride plastics are presented in review [9].

Radiation-coagulation purification. The method consists of the irradiation of wastes with the subsequent precipitation of the products of radiolytic transformations of pollutants by means of the addition of harmless coagulant [15, 16]. For instance, it was obtained [15, 16] that radiation-coagulation method can be used for the purification of wastes from polyoxyethylene *n*-nonyl phenyl ether $C_9H_{19}C_6H_4(C_2H_4O)_nH$ (n = 7·5, 10, 20 and 80), which is non-ionic detergent not decomposing biologically, and from polyvinyl alcohol with application of $Fe_2(SO_4)_3$ as a coagulant.

Radiation-ozone treatment. The method is based on the combination of ozonation (i.e. treatment by ozone) and irradiation (see reviews [15, 16]). Ozonation is used for the purification of some wastes since ozone is powerful oxidizing agent. However its action is selective; it is limited by organic insaturated molecules. The irradiation of aqueous wastes saturated with ozone intensifies the effect of ozonation. Combined action of ozone and ionizing radiation on the wastes causes the synergistic effect; the result of their joint action is higher than the sum of the results of separate actions of ozone and ionizing radiation.

From review [16] it follows that the synergistic effect under consideration takes place in the case of aqueous solution of phenol, ethylene glycol and oxalic acid. Radiation-chemical yields of decomposition of these compounds in the presence of ozone achieve 40, 117 and 87 (calculating on carbon), respectively. Such high yields testify that chain reactions occur.

The high yields of decomposition cause the good perspectives for practical application of radiation-ozone method. As it follows from calculations performed in work [16] the power of radiation source (provided the coefficient of using the radiation is 1) for the purification of waste containing 50 mg/l phenol (calculating on carbon) at output 5000 m³ a day should be only 35 kW.

Another example is the radiation-flotation-ozonation purification of wastes from surfactants (for instance, from nekal, which is a mixture of isomeric butylnaphthalene sulfonates and is used as an emulgator in the production of synthetic rubber) [21–23]. Several years ago it was found (see [24]), that radiation-flotation method was effective for the decomposition of nekal in the foamed state. To intensify this process it was proposed [21–23] to combine the irradiation of the foam containing nekal with the ozonation. In respective installation one part of electron beam is used for the synthesis of ozone which is directed then for the ozonation; another part is used for tertiary treatment of wastes after ozone-flotation purification. The source of electrons is the accelerator of ELV-3 type (energy is 0·7 MeV, power is $\approx$ 50 kW). The method is checked at Voronezh plant of synthetic rubber [22, 23].

Therefore it is possible to conclude that the combined technique of irradiation with conventional method increases considerably the probability of large scale application of ionizing radiation for the purification of wastes.

Radiation treatment of sewage sludge

During settling the wastes (it is the first stage of their purification) the sludges consisting of insoluble solid particles are formed. The composition of the sludges is 5–8% of solids and 92–95% of water. During biological purification of wastes the sludges of other type are formed at the stage of aeration; considerable fraction of them are the microorganisms. The sludges can be used as soil conditioner and supplement to animal feed. However, they are infected by bacteria, parasites and viruses and can contain toxic compounds. Obviously, the sludges before using should be disinfected.

Many researchers have shown (see [3, 4]) that the problem of disinfection of the sludges can be solved successfully by radiation methods. Besides, often the irradiation accelerates the filtration and the sedimentation of deposits and as a consequence the procedure of dewatering.

The disinfection of the sludges is achieved at doses 3–10 kGy (in dependence of the infection level) [8]. At present the technology of radiation disinfection of the sludges is developed; in some cases it is realized in pilot scale. In respective plants both the sources of gamma radiation of ^{60}Co [25] or ^{137}Cs [26, 27] and electron accelerators [8, 28, 29] are used. Let us consider in some detail the pilot plant designed in the Soviet Union [5, 8].

The scheme of the pilot plant is shown in Fig. 4. In this figure 1 is a mixer for raw sludge, 2 is a regulating vessel,

3 is a equipment for electron irradiation, 4 is an accelerator. It is the accelerator of ILU-6 type (energy of electrons is 2 MeV, mean power is 20 kW). The sludge irradiated at dose 10 kGy is moved through vessels 5 and 6 to a press-filter 7 where it is dewatered up to moisture content of 50–60% and then to a dryer 8 in which the sludge is dried to moisture content of 15%.

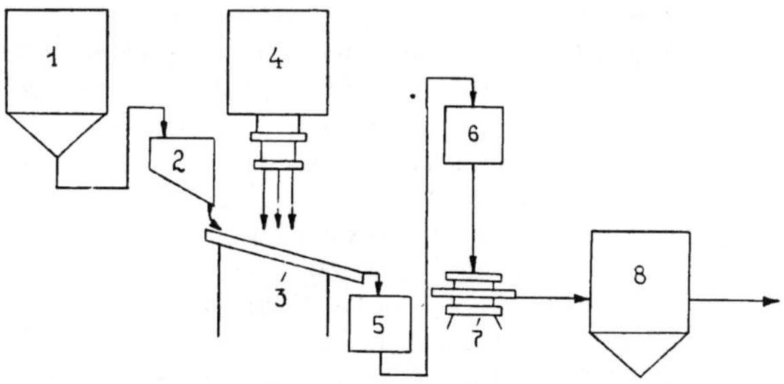

Fig. 4. Scheme of pilot plant for radiation treatment of sludges

The output of the installation is 2000 tons of dried sludge a year.

The analysis of the results obtained in pilot scale experiments shows that the perspectives of the large scale application of radiation treatment of sewage sludges are promising.

The problem which is close to the problem of disinfection of sewage sludges is the problem of disinfection of wastes of livestock breeding farms. For the solution of last problem the possibility of using the ionizing radiation is studied thoroughly in the U.S.S.R.

The problem mentioned is very complex. It is due to the high level of the infection and the high content of organic pollutants in the wastes of live-stock breeding farms. The organic compounds presenting in the wastes scavenge the free radicals formed from water radiolysis and as a consequence protect bacteria, parasites and viruses from indirect action of ionizing radiation. Because of it the doses required for the disinfection of such wastes are high (25 kGy and even higher) [30]. From economical point of view it is very important to reduce the dose required for the disinfection. The reduction can be achieved by different methods. In this connection the combination of radiation treatment with using the conventional disinfecting agents (chlorine, ozone, ammonia etc.) or sensitizing additions (for instance, KCl) is useful [30]. In review [30] it was noted that the addition of 0·2% of ammonia into liquid manure waste decreases the dose required for 100% disinfection to 5 kGy.

Radiation purification of exhaust gases

Another important problem of environmental conservation is the purification of exhaust gases from harmful impurities. Such impurities are sulfur dioxide, nitrogen oxides NO_x, carbon oxide, hydrocarbons etc. One of the promising ways of the solution of this problem is the application of ionizing radiation.

In the Soviet Union the radiation-catalytic method of purification of industrial gases from SO_2 was developed [6]. It consists of the passage of the gas through the layer of diluted aqueous solution of sulfuric acid containing $5 \cdot 10^{-3}$ mole/l $MnSO_4$ and simultaneous gamma irradiadiation. The radiolysis products of this solution oxidize Mn(II) to Mn(III); the latter acts as effective carrier of oxygen to SO_2 via the scheme including the reactions [6]:

$$Mn(III) + O_2 \rightarrow Mn(III) \cdot O_2 , \quad (5)$$

$$Mn(III) \cdot O_2 + SO_2 \rightarrow Mn(III) \cdot O + SO_3 , \quad (6)$$

$$Mn(III) \cdot O + SO_2 \rightarrow Mn(III) + SO_3 , \quad (7)$$

$$SO_3 + H_2O \rightarrow H_2SO_4 . \quad (8)$$

The process, as it is seen, is chain, and the formation of sulfuric acid occurs simultaneously with the purification.

Radiation-chemical yield of the oxidation depends on dose rate and oxygen concentration [6]. The yield is increased with the growth of oxygen concentration and with the decrease of dose rate. Maximum value of the yield is 10^4 molec./100 eV. The process is effective at sulfuric acid concentration $< 7{\cdot}5$ M. At higher concentration the oxidation yield is sharply reduced (because of the decrease of oxygen solubility).

The oxidation of SO_2 is inhibited by different impurities: phenol, Fe(II) ions etc. Because of it the method can not be used for the purification of flue gases of coal burning power plants [6]. These gases contain phenol which should be removed preliminary.

Obviously, in this case the expences for the purification are considerably higher. However, the radiation-catalytic method is perspective for the purification of exhaust gases of the plants producing sulfuric acid.

During several years the radiation methods of the purification of exhaust gases of iron-ore sintering furnaces, coal burning power plants etc. are developed in Japan and U.S.A. [31–35]. The purpose is the simultaneous removal of SO_2 and NO_x. In this direction some successes were achieved. In particular, it was found that the removal of SO_2 and NO_x was intensified noticeably by the addition of ammonia into the gas under the purification. In this case the flue gas is purified from SO_2 and NO_x by more than 80–90% at doses 10–15 kGy. Besides, SO_2 and NO_x are removed in the form of solid powder (its composition is the mixture of $(NH_4)_2SO_4$ and $(NH_4)_2SO_4 \cdot 2\,NH_4NO_3$) which can be used as a fertilizer. In 1978 this process was realized in Japan in large pilot scale [33–35]. According to [35], in U.S.A. the large pilot plant for the purification of the flue gases of coal burning plants by radiation method (with addition of ammonia) is under construction. The maximum feed rate of gas is planned to be 24000 m^3 an hour (under normal conditions).

The combined method of the purification of such gases is described in paper [36]. It consists of three stages:

1) the partial removal of SO_2 in the form of calcium sulfite and sulfate with using the lime;
2) irradiation by fast electrons (at this stage nitrogen oxides and SO_2 are oxidized to nitric and sulfuric acids, respectively; the acids react with lime);
3) the removal of powdered salts by means of fabric filters.

At present the pilot plant for checking the technological details of the method is under construction [36]. Radiation source is electron accelerator with two accelerating tubes (energy is 0·75 MeV, power is 40 kW).

The investigations in the field of the radiation purification of exhaust gases from SO_2 and NO_x are carried out also in the Soviet Union (see e.g. [37]). Now they are at the stage of laboratory experiments.

The wide large scale application of radiation methods for the purification of flue gases of coal burning power plants is closely connected with the development of accelerators technique. According to [36], the radiation power, required to purify the flue gases of coal burning power plant with electrical power of 500 MW (provided the dose for the purification is 13 kGy and coefficient of using the radiation is 0·8), is equal to 12·3 MW. Obviously, the radiation method can be realized if the electron accelerators of megawatt power will be available.

Conclusion

The consideration of current status of environmental applications of radiation chemistry and radiation technology shows that in many cases the radiation methods are promising for the purification of wastes and exhaust gases and for the treatment of sewage sludges. However, at present in this field there are no any large scale industrial processes with using the ionizing radiation; even the examples of pilot plants are rare. The elimination of this discrepancy is the important task of specialists on radiation chemistry and radiation technology.

Received September 27, 1985

Accepted in revised form April 29, 1986

References

[1] *P. I. Dolin, V. N. Shubin, S. A. Brusentseva,* Radiation Purification of Water, Nauka. Moscow 1973 (in Russian)

[2] *V. N. Shubin, S. A. Brusentseva, G. K. Nikonorova,* Radiation-Polymerization Purification of Wastes, Atomizdat. Moscow 1979 (in Russian)

[3] *N. A. Vysotskaya, A. F. Rekasheva, L. G. Shevchuk,* Radiation Treatment of Sewage Sludges, Tekhnika. Kiev 1980 (in Russian)

[4] *A. M. Dmitriev, V. N. Kalinin, V. S. Vetrov,* Radiation Treatment of Wastes of Live-Stock Breeding Farms, Urozhai. Minsk 1981 (in Russian)

[5] *V. S. Vetrov, N. A. Vysotskaya, A. M. Dmitriev, V. N. Kalinin, L. G. Shevchuk,* Radiation Treatment of Wastes for Agricultural Using, Energoatomizdat. Moscow 1984 (in Russian)

[6] *R. B. Baranova, L. T. Bugaenko, V. M. Byakov, V. I. Lazarev, E. P. Petryaev,* Exhaust Gases and Their Radiation-Chemical Purification, Energoizdat. Moscow 1981 (in Russian)

[7] *R. V. Dzhagatspanyan, V. A. Gol'din, M. T. Filippov, E. L. Mendel'son,* Technology of Radiation Purification of Wastes, Energoizdat. Moscow 1981 (in Russian)

[8] *N. A. Vysotskaya,* in: The Role of Chemistry in Environmental Conservation, Naukova Dumka. Kiev 1983, pp. 205–217 (in Russian)

[9] *S. A. Brusentseva, P. I. Dolin, V. E. Kazarinov, V. N. Shubin,* in: The Role of Chemistry in Environmental Conservation. Naukova Dumka. Kiev 1983, pp. 217–230 (in Russian)

[10] *A. M. Dmitriev,* in: The Role of Chemistry in Environmental Conservation, Naukova Dumka. Kiev 1983, pp. 230–242 (in Russian)

[11] *A. K. Pikaev, V. N. Shubin,* Radiation Phys. and Chem. **24** (1984) 77

[12] *V. N. Shubin, S. A. Brusentseva, N. A. Vysotskaya,* High Energy Chemistry **19** (1985) 338 (in Russian)

[13] Radiation Phys. and Chem. **24** (1984) 1

[14] *M. Washino,* Radiation Phys. and Chem. **18** (1981) 383

[15] *M. Takehisa, A. Sakumoto,* in: Industrial Application of Radioisotopes and Radiation Technology, IAEA. Vienna 1982, pp. 217 to 233

[16] *A. Sakumoto, T. Miyata,* Radiation Phys. and Chem. **24** (1984) 99

[17] *D. D. Woodbridge, P. C. Cooper, A. G. van der Berg, H. D. Musselman, H. N. Lowe,* in: Radiation for Clean Environment. IAEA. Vienna 1975, pp. 169–188

[18] *E. Lyu, I. E. Makarov, Yu. I. Kuryatnikov, R. V. Mazanko, R. V. Dzhagatspanyan, A. K. Pikaev,* Radiation Phys. and Chem. **22** (1983) 503

[19] *E. Lyu, I. E. Makarov, A. K. Pikaev,* High Energy Chemistry **17** (1983) 50 (in Russian)

[20] *E. Lyu, I. E. Makarov, A. K. Pikaev,* in: Proc. 5th Tihany Symposium on Radiation Chemistry, Akadémiai Kiadó. Budapest 1983, vol. 1, pp. 215–220

[21] *S. P. Buslaeva, Yu. A. Panin, L. B. Upadyshev, B. M. Vanyushkin, L. M. Makarochkina, V. F. Karpukhin, A. F. Nazarov, I. P. Konovalov,* in: All-Union Conference on Pure and Applied Radiation Chemistry (Obninsk, October 16–18, 1984), edit. by *A. K. Pikaev,* Nauka. Moscow 1984, p. 70 (in Russian)

[22] *S. P. Buslaeva, L. M. Makarochkina, A. Yu. Panin, I. A. Sobolev, D. V. Semenov, Yu. I. Kuryatnikov, L. B. Upadyshev,* in: The Abstracts of Papers of 2nd All-Union Meeting on Radiation Technology and Construction of Radiation Installations (September 26 to 30, 1983) VNIIRT. Moscow 1983, p. 41 (in Russian)

[23] *A. F. Nazarov, V. N. Semechev, E. V. Vorob'ev, Yu. A. Panin, S. P. Buslaeva, L. B. Upadyshev, L. M. Makarochkina,* in: The Abstracts of Papers of 2nd All-Union Meeting on Radiation Technology and Construction of Radiation Installations (September 26 to 30, 1983), VNIIRT. Moscow 1983, p. 42 (in Russian)

[24] *R. V. Dzhagatspanyan, V. I. Kosorotov, M. R. Filippov,* Introduction to Radiation-Chemical Technology, Atomizdat. Moscow 1979 (in Russian)

[25] *T. Lessel, A. Suess,* Radiation Phys. and Chem. **24** (1984) 3

[26] *J. S. Sivinski,* Radiation Phys. and Chem. **22** (1983) 99

[27] *J. S. Sivinski, S. Ahlstrom,* Radiation Phys. and Chem. **24** (1984) 17

[28] *J. G. Trump, E. M. Merrill, K. A. Wright,* Radiation Phys. and Chem. **24** (1984) 55

[29] *M. R. Cleland, R. A. Fernald, S. R. Maloof,* Radiation Phys. and Chem. **24** (1984) 179

[30] *V. N. Shubin,* High Energy Chemistry **17** (1983) 394 (in Russian)

[31] *S. Machi, O. Tokunaga, K. Nishimura,* Radiation Phys. and Chem. **9** (1977) 371

[32] *K. Kawamura, T. Katayama, K. Kawamura,* Radiation Phys. and Chem. **18** (1981) 389

[33] *K. Kawamura, S. Aoki, W. Kawakami, S. Hashimoto, S. Machi,* in: Radiation for Clean Environment, IAEA. Vienna 1975, pp. 621 to 631

[34] *K. Kawamura, V. H. Shui,* in: Industrial Application of Radioisotopes and Radiation Technology, IAEA. Vienna 1982, pp. 197 to 215

[35] *K. Kawamura, V. H. Shui,* Radiation Phys. and Chem. **24** (1984) 117

[36] *D. J. Helfritch, P. L. Feldman,* Radiation Phys. and Chem. **24** (1984) 129

[37] *E. V. Belousova, B. G. Dzantiev, I. K. Larin, S. U. Pavlova, D. N. Sapozhnikov, A. P. Shvedchikov,* ZfI-Mitteilungen **97** (1984) 381

Isotopenpraxis 22 (1986) 12, S. 444 bis 446

IR-spektroskopische und mechanistische Untersuchungen an Technetiumkomplexen unter Einsatz von ^{15}N

B. Lorenz, K. Schmidt, L. Kaden, M. Wahren

(Akademie der Wissenschaften der DDR, Zentralinstitut für Isotopen- und Strahlenforschung, Leipzig)[1]

Die Bildung von Nitrido- und Thionitrosylkomplexen des Technetiums wurde unter Einsatz von ^{15}N-markierten Verbindungen untersucht. Trotz völlig analoger Ausgangs- und Endprodukte weist der Reaktionsverlauf bei Technetium- und Rheniumkomplexen wesentliche Unterschiede auf. Eine neue glatte Synthese von Nitridotechnetium(V)-Komplexen aus Komplexen des Tc(III) und Azid wurde gefunden. Die ^{15}N-Isotopensubstitution ermöglicht eine eindeutige Zuordnung charakteristischer IR-Banden der untersuchten Komplexe.

The formation of Technetium nitrido- and thionitrosyl-complexes was investigated by means of ^{15}N labelled compounds. The course of the reaction is rather different with the elements technetium and rhenium in spite of quite analogous starting materials and products. A new feasible synthesis of nitridotechnetium (V) complexes was found from complexes of Tc(III) and azide. By ^{15}N-substitution an unambiguous assignment of characteristic infrared bonds in the complexes was achieved.

Keywords

azides; chemical preparation; hydrazine; infrared spectra; nitrogen 15; technetium 99; technetium complexes

Durch Einsatz von ^{15}N-Verbindungen lassen sich komplexe Reaktionsabläufe untersuchen. Speziell können charakteristische IR-Banden von Zwischen- und Reaktionsprodukten durch Isotopensubstitution eindeutig zugeordnet werden.

Für die Darstellung von Technetiumnitridokomplexen sind gegenwärtig zwei Methoden bekannt. Aus Pertechnetat, tertiärem Phosphan und Hydrazinsalz in ethanolischer HCl bilden sich Nitridokomplexe des Tc(V) [1] und aus Pertechnetat, konz. HCl und Natriumazid entsteht $(TcNCl_4)^-$ mit Tc(VI) [2].

Von *Chatt* [3] wurde 1969 die Reaktion von Perrhenat mit Hydrazin und tertiären Phosphanen zu Rhenium(V)-nitridokomplexen beschrieben, ihr Mechanismus untersucht und dabei festgestellt, daß die Reduktion des Re(VII) durch überschüssiges Phosphan erfolgt, wobei sich Phosphanoxid bildet. Wir fanden nun, daß sich im Falle des Tc kein Phosphanoxid, sondern gasförmiger Stickstoff bildet [4].

Seine quantitative Bestimmung durch Ausführen der Reaktion unter Ar wurde bei der mehrstufigen Umsetzung durch eindringende Spuren von Luft gestört, gelang aber durch Einsatz von ^{15}N-markiertem Hydrazinsalz, wodurch der Stickstoff aus der Reaktion an ^{15}N angereichert war. Dabei wurde mit Perrhenat nahezu kein Stickstoff gebildet (7% der stöchiometrisch nach Gleichung (2) zu erwartenden Menge) in Übereinstimmung mit dem von *Chatt* und Mitarb. [3] vorgeschlagenen Reaktionsmechanismus und der Bruttoreaktionslgeichung (1)

$$\begin{aligned} &NaReO_4 + N_2H_6Cl_2 + 2\,HCl + 4\,PPh_3 \\ &\rightarrow ReNCl_2(Ph_3P)_2 + 2\,Ph_3PO + NaCl + NH_4Cl \\ &+ 2\,H_2O\,. \end{aligned} \tag{1}$$

Überraschend verläuft jedoch bei Einsatz von Pertechnetat die Synthese zu 100% nach Bruttoreaktionsgleichung (2), wobei Hydrazin als Reduktionsmittel für Tc(VII) wirkt:

$$\begin{aligned} &NaTcO_4 + 2\,N_2H_6Cl_2 + 2\,Ph_3P \rightarrow TcNCl_2(Ph_3P)_2 \\ &+ N_2 + NaCl + NH_4Cl + 4\,H_2O\,. \end{aligned} \tag{2}$$

[1]) Anschrift: DDR-7050 Leipzig, Permoserstr. 15

Für den Ablauf dieser komplexen Reaktion läßt sich annehmen, daß aus Pertechnetat unter Oxidation von Hydrazin ein Chlorophosphankomplex des dreiwertigen Technetiums gebildet wird, der mit Azid unter Substitution von Chlor und anschließender intramolekularer Redoxreaktion gemäß (3)

$$L_nTcN_3 \rightarrow L_nTc \equiv N + N_2 \tag{3}$$

in den Nitridokomplex des Tc(V) übergehen könnte (L = Ligand; n = Anzahl der Liganden).

Es ist seit langem bekannt [5, 6], daß bestimmte Oxidationsmittel, darunter $KMnO_4$ in saurer Lösung, Hydrazin unter zumindest teilweiser intermediärer Bildung von Azid nach (4) oxidieren

$$2\,N_2H_4 \xrightarrow[-2\,H_2]{Ox.} 2\,(N_2H_2) \rightarrow (N_4H_4) \rightarrow NH_4^+N_3^-\,. \tag{4}$$

Diese Reaktion ist auch zumindest teilweise der Grund dafür, daß ^{15}N-Hydrazin mit nicht statistischer Isotopenverteilung mit $KMnO_4$ nur in alkalischer Lösung zu N_2 mit gleicher Isotopenverteilung oxidiert wird; in saurer Lösung wird N_2 erhalten, dessen Isotopenverteilung in Richtung auf die Einstellung des Isotopenaustauschgleichgewichts verschoben ist [7, 8]. Sowohl $NaTcO_4$ als auch $NaReO_4$ verhalten sich analog, wie von uns gefunden wurde, daher wird intermediär Azid in der Reaktionslösung mit großer Wahrscheinlichkeit vorhanden sein.

Der weiter oben postulierte Teilschritt, die Reaktion von Chlorophosphantechnetium(III)-Komplexen mit Azid, konnte ebenfalls verifiziert werden. Die entsprechende Reaktion mit Rheniumkomplexen war bereits bekannt. Sie verläuft jedoch nicht glatt und nur mit relativ geringer Ausbeute an Rheniumnitridokomplexen [3].

Dagegen reagieren Technetium(III)-Komplexe des Typs $TcCl_3L_3$ (L = Dimethylphenyl- bzw. Diethylphenylphosphan) mit Natriumazid in siedendem Ethanol in glatter Reaktion und sehr guten Ausbeuten zu Nitridophosphankomplexen ($TcCl_2NL_3$) des Technetium(V). Damit wurde neben dem Ligandenaustauschprinzip [1a] ein weiterer und merklich effektiverer Syntheseweg für diesen Komplextyp gefunden. Mit der Methode lassen sich durch Einsatz von ^{15}N-Natriumazid (^{15}N = 95 At.%) problemlos die isotop-substituierten Verbindungen darstellen.

Im IR-Spektrum ändert sich die Lage der Valenzschwingung Tc $\equiv$ N bei der Dimethylphenylphosphanverbindung von 1048 cm^{-1} [1a] auf 1019 cm^{-1} und bei der Diethylphenylphosphanverbindung von 1038 cm^{-1} auf 1006 cm^{-1}. Beim letztgenannten Komplex wurde durch die Isotopensubstitution eine deutliche Separierung der Banden und somit eine eindeutige Zuordnung der Valenzschwingung Tc$\equiv$N möglich. Des weiteren ist der Komplex ($Tc^{15}NCl_2(Me_2PhP)_3$) ein idealer Ausgangsstoff für die Synthese von isotop-markierten Thionitrosylkomplexen des Technetiums, deren erste Vertreter von uns kürzlich in unmarkierter Form synthetisiert wurden [9].

Die charakteristischen, intensiven ν_{NS} Absorptionsbanden in den IR-Spektren verschieben sich beim Technetium(I)-Komplex von 1177 cm^{-1} auf 1147 cm^{-1} und beim Technetium(II)-Komplex von 1240 cm^{-1} auf 1206 cm^{-1}. Eine beim unmarkierten Technetium(II)-Komplex neben der bei 1240 cm^{-1} gelegenen gleichfalls neu auftretenden Bande bei 1147 cm^{-1} bleibt bei Isotopensubstitution lagekonstant und kann bisher nicht zugeordnet werden. Interessanterweise tritt eine derartige Bande auch beim Dichlorobis(diethyldithiocarbamato)-thionitrosyltechnetium-(III)chelat bei 1150 cm^{-1} auf [10].

Im Gegensatz zu den genannten Technetium(III)-Verbindungen zeigen Technetium(IV)-Verbindungen einen anderen Reaktionsverlauf. Setzt man Technetium(IV)-Verbindungen des Typs $TcCl_4L_2$ mit Natriumazid in siedendem Ethanol um, so tritt neben Gasentwicklung ein deutlicher Farbwechsel von grün nach braun auf. Das aus der Lösung nach Einengen und Ausfällen mit Ether gewonnene mikrokristalline, braune Produkt ist jedoch nicht der erwartete monomere Technetium(VI)-Komplex ($TcNCl_3L_2$).

Im IR-Spektrum fehlt die für terminale M $\equiv$ N-Gruppen charakteristische Bande im Gebiet zwischen 1000 bis 1100 cm^{-1} [11, 2], und auch das EPR-Spektrum (CH_2Cl_2) ist wenig intensiv und untypisch für eine d^1-Elektronenkonfiguration [2].

Das IR-Spektrum zeigt ferner eine deutliche Intensitätserhöhung der bei 1137 und 1110 cm^{-1} liegenden Banden. Die Bande bei 1110 cm^{-1} verändert ihre Lage infolge Isotopensubstitution durch ^{15}N nur unwesentlich (1110 $\rightarrow$ 1097 cm^{-1}).

Indes ist bemerkenswert, daß die sich bei längerem Stehen einer etherischen Lösung von $TcNCl_2L_3$ (L = Me_2PhP) über eine rotviolette Zwischenstufe bildende braune Verbindung ein identisches IR-Spektrum ergibt und auch nahezu gleiche elementaranalytische Werte aufweist. Die Elementaranalyse ergab für Tc 22,7%; N 2,8% C 34,1%. Die bisherigen Ergebnisse erlauben z. Zt. noch keine eindeutige Charakterisierung der Verbindung.

Der Umsatz von NH_4TcO_4 mit ^{15}N-Natriumazid in konz. HCl lieferte in Analogie zur unmarkierten Verbindung nach *Baldas* et al. [2] problemlos die isotop-markierte Verbindung Tetraphenylarsoniumtetrachloronitridotechnetat (VI) ($AsPh_4$) ($Tc^{15}NCl_4$). Im IR-Spektrum ist die $\nu_{Tc\equiv N}$-Bande gegenüber der unmarkierten Veibindung ($\nu_{Tc\equiv N}$ = 1076 cm^{-1}) um 28 cm^{-1} verschoben und befindet sich bei 1048 cm^{-1}. Die unmarkierte Verbindung in Form des Isomeren ^{99m}Tc kann, wie kürzlich mitgeteilt, als Ausgangsstoff für viele Markierungssynthesen in der Nuklearmedizin benutzt werden [12]. Wir verwenden die ^{15}N-isotopmarkierte Verbindung zur Darstellung des ^{15}N-Nitrido-bis(diethyldithiocarbamato)technetium(V) durch Reaktion mit Natriumdiethyldithiocarbamat in Acetonitril. Mit Hilfe des IR-Spektrums dieser Verbindung gelang es, die von uns fälschlicherweise als $\nu_{C=S}$ zugeordnete Valenzschwingung bei 1074 cm^{-1} [1] zweifelsfrei als Valenzschwingung Tc $\equiv$ ^{14}N zuzuordnen, denn diese Bande verschiebt sich bei ^{15}N-Markierung auf 1033 cm^{-1}.

In Tab. 1 sind sämtliche dargestellte ^{15}N-isotopmarkierte Verbindungen aufgeführt.

Für die Bildung des $(TcNCl_4)^-$-Ionswurden intermediär auftretende Technetium(IV)-Spezies und deren Weiterreaktion mit Azid in Betracht gezogen [2]. Um diese Annahme zu prüfen, setzten wir K_2TcCl_6 in konz. HCl mit Natriumazid um. Es gelang uns jedoch nicht, das erwartete Nitridotechnetat(VI) zu isolieren. Statt dessen wurde immer nur das hellgelbe Tetraphenylarsoniumhexachlorotechnetat(IV) erhalten und anhand seines IR-Spektrums (charakteristische Banden bei 1086 cm^{-1}, 470 cm^{-1}, 454 cm^{-1} [13]) identifiziert.

Zusammenfassend kann festgestellt werden, daß Azidokomplexe des Technetium(III) zumindest bei der Darstellung von Phosphannitridokomplexen des Technetium(V) als Zwischenprodukte eine dominierende Rolle spielen. Unklar ist z. Zt. noch der Reaktionsverlauf beim Umsatz von Natriumazid mit Technetium(IV)-phosphankomplexen, wogegen Tc(IV)-hexachlorokomplexe bei der Bildung von $(TcNCl_4)^-$ keine Bedeutung besitzen. Mit der Darstellung von ^{15}N-isotopmarkierten Verbindungen wurde eine eindeutige IR-spektroskopische Zuordnung einiger Banden ermöglicht.

Experimentelles

Trichlorotrisdimethylphenylphosphantechnetium(III) und Tetrachlorobisdimethylphenylphosphantechnetium(IV) sowie die entsprechenden Diethylphenylphosphanderivate und K_2TcCl_6 wurden nach der Literatur [15, 16] dargestellt.

Tab. 1. Valenzschwingungen einiger Nitrido- und Thionitrosylkomplexe des Technetiums, Nitridokomplexe terminaler Typ Tc $\equiv$ N

Nr.	Verbindung	$\nu_{Tc\equiv ^{14}N}$ (cm^{-1})	Lit.	$\nu_{Tc\equiv ^{14}N}$ (cm^{-1})	Lit.	$\Delta\nu$ (cm^{-1})
1	$TcCl_2N$ $(Me_2PPh)_3$	1048	[1a]	1019		29
2	$TcCl_2N$ $(Et_2PPh)_3$	1038	[1a]	1006		32
3	$(AsPh)_4$ $(TcNCl_4)$	1076	[2]	1048		28
4	TcN $(dtc)_2$	1074	[1a]	1033		41
5	TcNpc	1078	[14]	1049	[14]	29
	Thionitrosyle	$\nu_{^{14}NS}$		$\nu_{^{15}NS}$		$\Delta\nu$
6	$TcCl_2NS$ $(MePPh)_3$	1177	[9]	1147		30
7	$TcCl_3NS$ $(Me_2PPh)_2$	1240	[9]	1206		34

Abkürzungen:

Me = Methyl — dtc = Diethyldithiocarbamat

Et = Ethyl

Ph = Phenyl — pc = Phthalocyanin

^{15}N-Natriumazid (95 At.%) war ein Produkt der Firma Isocommerz GmbH. ((Ph_4As)) ($^{15}NCl_4Tc$) wurde nach [2] aus NH_4TcO_4 und ^{15}N-Natriumazid dargestellt und daraus durch Umsetzung mit Natriumdiethyldithiocarbamat in Acetonitril ^{15}N-Nitridobis(diethyldithiocarbamato)technetium(V) gewonnen. Das nachfolgend beschriebene ^{15}N-Nitridodichloro-tris(dimethylphenylphosphan)technetium-(V) diente als Ausgangsprodukt für die Darstellung der Thionitrosylkomplexe nach [9].

Die Aufnahme der IR-Spektren (KBr-Preßlinge) erfolgte am Gerät UR 10 (VEB Carl Zeiss Jena).

Reaktion von $NaTcO_4$ und $NaReO_4$ mit ^{15}N-Hydrazin

Es wurde in gleicher Weise, wie in [8] angegeben, gearbeitet. Die verwendete Testmischung aus Hydrazinsulfat-$^{15}N_2$ und Hydrazinsulfat mit natürlicher isotoper Zusammensetzung lieferte bei der Oxidation mit $KI/I_2/HCO_3^-$ oder alkalischer $KMnO_4$-Lösung die in nachstehender Tab. 2 als „Standard" bezeichnete isotope Zusammensetzung des Stickstoffs.

Tab. 2. Umsetzung von Hydrazinsulfat mit $NaTcO_4$ und $NaReO_4$

Probe	$^{14}N_2$	$^{14}N^{15}N$	$^{15}N_2$	^{15}N-Häufigkeit [Atom-%]
Standard	49,1	4,3	46,5	48,7
$NaTcO_4$, 2 N NaOH 40°	49,4	4,4	46,2	48,4
$NaReO_4$, 2 N NaOH 40°	49,1	4,3	46,5	48,6
$NaTcO_4$, 2 N H_2SO_4	48,6	6,3	45,1	48,3
$NaReO_4$, 2 N H_2SO_4	49,1	4,9	46,0	48,5
$KMnO_4$, 2 N H_2SO_4[1])	48,0	8,1	43,8	47,9

[1]) Werte aus [8] zum Vergleich mit ähnlicher Testmischung

^{15}N-Nitridodichloro-tris(dimethylphenylphosphan) technetium (V)

365 mg (0,57 mmol) $TcCl_3$ $(Me_2PhP)_3$ werden unter Ar in 25 ml abs. Ethanol gelöst und hierzu 44,2 mg (0,65 mmol) ^{15}N-Natriumazid in fester Form zugegeben. Die Farbe der roten Lösung ändert sich in wenigen Minuten nach gelb und Gasentwicklung setzt ein. Die Lösung wird 2 h am Rückfluß gekocht, vom ausgefallenen NaCl abdekantiert und auf ca. 1/3 des Ausgangsvolumens im Vakuum eingeengt. Es resultieren gelbe Kristalle, die mit kaltem Ethanol gewaschen und aus Aceton umkristallisiert werden.

Fp: 170 °C Ausbeute: 290 mg (84%)

Eingegangen am 2. 6. 1986

Literatur

[1] a) *L. Kaden, B. Lorenz, K. Schmidt, H. Sprinz, M. Wahren*, Isotopenpraxis **17** (1981) 174
b) *J. Baldas, J. Bonnyman, P. M. Pojer, G. A. Williams, M. F. Mackay*, J. chem. Soc., Dalton Trans. (1981) 1798

[2] *J. Baldas, J. F. Boas, J. Bonnyman, G. A. Williams*, J. chem. Soc., Dalton Trans. (1984) 2395

[3] a) *J. Chatt, C. D. Falk, G. J. Leigh, R. J. Paske*, J. chem. Soc. (London) (A) (1969) 2288
b) *J. Chatt, J. R. Dilworth, G. J. Leigh*, J. chem. Soc. (London) (A) (1970) 2239

[4] *M. Wahren, B. Bayerl, B. Lorenz, L. Kaden, B. Plarre, K. Schmidt*, Z. Chem. **23** (1983) 71 (Kurzreferat)

[5] *L. F. Audrieth, B. A. Ogg*, The Chemistry of Hydrazine. New York: John Wiley 1951, Kap. 6

[6] *A. Brown, W. C. E. Higginson*, J. chem. Soc., Dalton Trans. (1972) 166

[7] *W. C. E. Higginson, D. Sutton*, J. chem. Soc. (London) (1953) 1402

[8] *M. Wahren, B. Lorenz*, Isotopenpraxis **10** (1974) 273

[9] *L. Kaden, B. Lorenz, R. Kirmse, J. Stach, U. Abram*, Z. Chem. **25** (1985) 29

[10] *J. Baldas, J. Bonnyman, M. F. Mackay, G. A. Williams*, Aust. J. Chem. **37** (1984) 751

[11] a) *K. Dehnicke, J. Strähle*, Angew. Chem. **93** (1981) 451
b) *U. Abram, H. Spies, W. Görner, R. Kirmse, J. Stach*, Inorg. chim. Acta **109** (1985) L 9

[12] *J. Baldas, J. Bonnyman*, IAEA-CN-45/46, 179

[13] *G. Bandoli, D. A. Clemente, U. Mazzi, E. Roncari*, J. Chem. Soc., Dalton Trans. (1982) 1381

[14] *S. Rummel, M. Hermann, K. Schmidt*, Z. chem. **25** (1985) 152

[15] *U. Mazzi, G. De Paoli, P. Di Bernardo, L. Magon*, J. inorg. nuclear. Chem. **38** (1976) 721

[16] *J. Dalziel, N. S. Gill, R. S. Nyholm, R. D. Peacock*, J. chem. Soc. (London) (1958) 4012

Isotopenpraxis 22 (1986) 12, S. 446

Buchbesprechung

Leben wir in einem Eiszeitalter ?

Heinz Schulz, Berlin: VEB Deutscher Verlag der Wissenschaften 1985, 85 S., 75 Abb., 1 Tab., 11 Lit.; karton. M 24.—

Die Geschichte der Eiszeittheorien wird als Grundlage für das Verständnis der Gegenwart nach einigen einleitenden Bemerkungen in Form mehrerer Kapitel ausführlich behandelt, ehe der Autor versucht, die im Titel gestellte Frage zu beantworten. Diese Kapitel behandeln die Sintflut-Legende, die Rolle von Eis und Schnee im frühen Denken, die Entstehung moderner Eiszeittheorien, das jüngste Eiszeitalter und die Frage der Entstehung von Eiszeitaltersverhältnissen. In dem Teil über Gegenwart und Zukunft rückt dann naturgemäß die Kohlendioxid-Problematik in den Mittelpunkt, die wohl in entscheidendem Maße von den Prognosen für den erwartete Weltenergiebedarf geprägt wird. Er wird seit einigen Jahren schon kontinuierlich nach unten korrigiert. Deshalb verwundert es auch nicht, daß die Zeiten, in denen das eine oder andere Klimamodell Realität werden kann, zwischen 30 und 23000 Jahren liegen, woraus auch folgt, daß die gestellte Frage in absehbarer Zeit kaum mit ja oder nein zu beantworten sein wird.

Insgesamt ein gut aufgemachtes Buch, welches auch durch die Farbabbildungen einen soliden Eindruck hinterläßt und all denen empfohlen werden kann, die sich mit dieser Problematik noch nicht beschäftigt haben.

Um einen kleinen Schönheitsfehler kommt aber auch dieses Buch nicht herum; „Kohlendioxyd" statt „Kohlendioxid" und „Grad C" statt „K" sollten der Vergangenheit angehören.

P. Harting

Isotopenpraxis 22 (1986) 12, S. 447 bis 450

Effect of Ionizing Radiation on the Distribution of Selenium in Rats

S. Kossakowski, A. Grosicki, A. Dziura

(Veterinary Research Institute, Pulawy, Department of Radiobiology)[1])

Investigations were carried out on 350 Wistar rats irradiated with doses of 52, 103, 155 and 205 mC. One or five days after irradiation the rats were given intragastrically dose of $Na_2{}^{75}SeO_3$ *(0·2 cm³, 3·34 kBq). The radioactivity of organs and tissues at 3, 6 h and 1, 2, 4, 8, 14 days after isotope administration indicated that irradiated rats did not differ significantly from not irradiated rats in distribution of radioselenium. Moreover the differences were transient and did not correlate with the exposure doses.*

Die Untersuchungen wurden an 350 Wistar-Ratten ausgeführt, die mit Dosen von 52, 103, 155 und 205 mC bestrahlt wurden. Nach einem oder nach fünf Tagen der Bestrahlung wurden den Ratten intragastrisch Dosen von $Na_2{}^{75}SeO_3$ *(0,2 cm³, 3,34 kBq) verabreicht und die Verteilung der Radioaktivität in den Organen und Geweben nach 3 und 6 h und nach 1, 2, 4, 8, und 14 Tagen untersucht. Zwischen den bestrahlten und nicht-bestrahlten Ratten konnten keine signifikanten Unterschiede in der Verteilung des Selens gefunden werden; die Unterschiede waren vorübergehend und korrelierten nicht mit den Bestrahlungsdosen.*

Keywords

biological localization; biological radiation effects; distribution; ionizing radiations; irradiation; organs; rats; selenium 75; tissues; X radiation

Introduction

Ionization is the basic mechanism through which ionizing radiation affects a living body. Since water is present in nearly all living cells, water ionization is of key importance in the pathogenesis of postirradiation damage [1, 2, 3, 4].

Because of that much attention has been given to antioxidants in the prophylaxis and in the therapy of postirradiation damage. It was as early as in 1951 that *Bacq* et al. [5] pointed out the radioprotective significance of cysteamine and cystamine reacting with radicals. *Zalkin* et al. [6] reported selenium compounds to be more effective antioxidants than the sulphur-containing analogues. *Hollo* and *Zlatarov* [7] reported that all rats irradiated with lethal doses survived after being treated with selenium. Subsequently *Olcott* et al. [8] showed selenomethionine to be a more potent antioxidant than methionine. *Hamilton* and *Tappel* [9] found that kidney and liver selenoproteids have much stronger antioxidative properties than vitamin E and that the synergism between vitamin E and selenium is weak. *Diplock* [10] regards vitamin E and selenium as a complex system which protects the cells against the harmful effects of free radicals. *Shimazu* and *Tappel* [11] found that selenoamino acids reduce the rate at which free radicals are generated as a result of gamma radiation. Radioprotective effect of selenium has been emphasized by *Badiello* et al. [12] whereas selenomethionine by *Shamberger* [13]. Recently *Cekan* et al. [14] furnished evidence that selenium protected against radiation-induced developmental defects in mice.

It should be also stressed that, apart from its considerable physiological importance, selenium can have pathogenic effects [15, 16, 17] since when present in excess it causes selenotoxicoses and when deficient it is the cause of numerous ailments in humans [18, 19, 20] and animals [21, 22, 23, 24].

In the light of the data it was thought worthwhile to undertake investigations into the effect of different doses of ionizing radiation on the distribution of selenium in the animal organism.

Experimental

Experiments were performed on 350 male Wistar rats weighing (165 ± 18) g. The animals were on standard laboratory chow and were allowed free access to tap water throughout the study. The rats were divided into 5 groups, each group containing 70 animals. Group 1 was a non-irradiated control. Groups from 2 through 5 received the following doses of radiation: 52 mC — group 2, 103 mC — group 3, 155 mC — group 4, and 206 mC — group 5. All rats were administered orally 0,2 cm³ of a $Na_2{}^{75}SeO_3$ solution produced by the Nuclear Research Institute at Świerk. The rats were given the isotope either 1 or 5 days after irradiation. Subsequently the rats were sacrificed after 3 h, 6 h, 1, 2, 4, 8, and 14 days. The following organs were taken for radiometric examinations: stomach, small and large intestine (without content), liver, kidneys, lungs, muscles, spleen, blood, brain, testis, prostate gland, skin, and hair. The samples were homogenized and 1-gram amounts were placed in a aluminium tube and analysed for radioactivity.

Irradiated rats were observed for clinical symptoms and in some animals of each group counts of lymphocytes and leucocytes were made. The rats were irradiated with an X ray generating device ("Siemens Bombe") using the following parameters: 20 mA, 180 kV, dose strength 0·98 mC per min. The amount of irradiation dose was measured using the Röntgen-gamma dosimeter manufactured by VEB Robotron, DDR. Radioactivity of the samples was measured three times by means of an NaI/Tl crystal scintillation detector type SSU-70 manufactured by "Polon".

Results were computed as the per cent of the radioselenium dose given to the animals. The data were analysed statistically using *Student*'s t test and significance was judged at the 0·05 level.

[1]) Address: Partyzantów 57, Pl-24-100 Puławy, Poland

Tab. 1. Concentrations of ^{75}Se in organs and tissues of rats 1 d after irradiation [% × 10^{-1}]

No.	Organs and Tissues	3 h					6 h					1 d				
		C	52 mC	103 mC	155 mC	206 mC	C	52 mC	103 mC	155 mC	206 mC	C	52 mC	103 mC	155 mC	206 mC
1.	Stomach	3·6	7·7	4·0	4·8	5·5	4·0	5·3	4·2	6·0	5·2	4·7	5·1	4·9	5·7	4·5
2.	Small intestine	51·0	47·5	47·4	63·3	78·1	33·3	43·4	49·1	47·1	45·4	18·3	23·1	28·1	33·1	34·2
3.	Large intestine	1·1	1·6	0·9	1·2	0·8	2·9	2·2	2·5	2·0	1·9	3·6	4·8	4·0	4·9	4·1
4.	Liver	13·2	17·6	16·1	17·2	14·8	14·9	17·7	15·9	14·2	14·1	17·3	18·2	21·3	18·4	16·1
5.	Kidney	10·3	16·5	12·3	14·3	9·9	25·8	21·0	23·7	18·0	20·5	25·9	36·5	34·6	32·2	25·1
6.	Heart	3·2	2·7	1·8	2·3	1·7	3·8	3·6	3·6	3·4	3·1	3·7	5·2	4·4	4·0	3·3
7.	Lung	3·2	3·7	2·6	2·8	2·5	5·6	4·9	5·3	5·2	4·3	5·9	7·2	6·3	6·6	5·3
8.	Muscle	0·1	0·2	0·2	0·3	0·2	0·6	0·7	0·7	0·7	0·5	1·3	1·7	1·6	1·4	1·1
9.	Spleen	2·6	3·4	2·7	3·6	2·7	4·9	4·3	5·8	4·6	4·3	6·3	8·0	7·7	8·0	6·3
10.	Blood	9·3	8·3	6·1	8·3	6·3	13·9	11·2	12·7	10·4	9·5	9·1	12·6	10·8	9·6	8·8
11.	Brain	0·4	0·2	0·3	0·5	0·3	0·6	0·5	0·6	0·5	0·6	0·9	1·3	1·0	1·0	0·8
12.	Testicle	1·2	2·0	1·2	1·4	1·2	5·9	3·7	3·6	3·4	2·8	6·1	8·9	7·2	6·7	5·3
13.	Prostate	0·6	1·2	0·3	1·4	0·5	2·6	1·9	2·4	2·1	0·9	2·8	4·2	3·2	4·1	2·9
14.	Skin	0·1	0·8	0·4	0·5	0·3	0·7	1·2	1·1	1·2	0·8	2·6	3·6	3·1	2·8	2·2
15.	Hair	0·1	0·2	0·1	0·2	0·1	0·1	0·1	0·1	0·1	0·1	0·1	0·1	0·1	0·1	0·1

Tab. 2. Concentrations of ^{75}Se in organs and tissues of rats 5 d after irradiation [% × 10^{-1}]

No.	Organs and Tissues	3 h					6 h					1 d				
		C	52 mC	103 mC	155 mC	206 mC	C	52 mC	103 mC	155 mC	206 mC	C	52 mC	103 mC	155 mC	206 mC
1.	Stomach	3·6	5·3	4·1	4·5	4·0	4·0	4·0	4·4	3·7	4·4	4·7	4·7	4·5	4·6	5·1
2.	Small intestine	51·0	61·8	49·8	42·8	66·7	33·3	51·7	48·5	53·9	49·1	18·3	20·9	16·4	16·7	12·9
3.	Large intestine	1·1	1·3	1·1	1·2	1·4	2·9	2·9	2·7	2·0	2·2	3·6	3·8	3·1	3·6	4·2
4.	Liver	13·2	19·1	18·7	14·1	15·0	15·0	16·1	16·5	13·7	12·3	17·3	18·7	13·8	12·5	13·7
5.	Kidney	10·3	17·5	12·4	12 3	17·0	25·8	23·6	22·2	19·9	18·5	25·9	29·4	27·0	28·6	31·2
6.	Heart	3·2	2·0	2·3	2·5	3·2	3·8	3·7	4·1	3·1	2·9	3·7	4·2	3·6	3·5	4·1
7.	Lung	3·2	3·4	3·3	3·5	4·5	5·6	5·3	6·1	4·5	4·3	5·9	7·0	5·9	5·5	6·2
8.	Muscle	0·1	0·3	0·3	0·3	0·4	0·6	0·6	0·6	0·5	0·4	1·3	1·5	0·9	1·2	1·3
9.	Spleen	2·6	3·1	3·6	3·2	4·2	4·9	5·0	5·5	4·6	4·9	6·3	8·5	7·6	7·4	8·6
10.	Blood	9·3	8·0	8·3	8·8	11·1	13·9	13·0	13·2	10·4	8·6	9·1	10·5	9·1	8·2	9·1
11.	Brain	0·4	0·4	0·5	0·4	0·5	0·6	0·7	0·6	0·6	0·5	0·9	1·1	0·9	0·9	1·2
12.	Testicle	1·2	1·5	1·5	1·7	1·9	5·9	3·7	3·6	2·8	3·1	6·1	8·0	6·9	6·1	8·0
13.	Prostate	0·6	0·7	1·0	0·9	1·2	2·6	1·3	2·1	1·3	1·6	2·8	2·8	2·2	3·1	3·9
14.	Skin	0·1	0·6	0·4	0·4	0·4	0·7	1·1	0·8	0·6	0·7	2·6	2·9	2·1	2·5	2·6
15.	Hair	0·1	0·1	0·1	0·2	0·1	0·1	0·1	0·1	0·1	0·2	0·1	0·1	0·1	0·2	0·1

Results

Per cent selenium contents in the investigated organs are listed in Tab. 1 and 2. The data indicate that in the control rats which were given a single oral dose of ^{75}Se the selenium distribution was a two-phase process in the majority of organs and tissues. Initially, there was a phase of increase followed by that of elimination. The exception to that were the testis and the hair where the build-up of selenium was observed over the whole length of the experiment. Among the organs and tissues could be distinguished those in which peak selenium concentrations occured after 3 h (small intestine) or 6 h (heart, blood) and those in which radioselenium build-up persisted until the 1st day (liver, kidneys, lung, prostate, skin), until the 2nd day (stomach, large intestine, spleen, muscles) and until the 14th day (testis, hair, brain). The highest concentrations of the isotope were found after 3 h in the small intestine, liver, and kidneys; after 6 h in the small intestine, kidneys, and liver and after 2 to 14 days in the kidneys, liver, and small intestine. Of interest are the low concentrations of radioselenium in the muscle tissue and in the brain.

After irradiation the rats of groups 3—5 were apathetic, showed thirst and lack of appetite. The symptoms intensified along with increasing irradiation doses and persisted up to 24 h in rats of group 5. After 10 to 14 days rats of groups 4 and 5 developed diarrhoea. Haematological examinations of the irradiated rats showed the maximum decrease in lymphocyte number after 1 day from 65% (group 3) to 75% (group 5), and maximum decrease in leukocyte number after 5 days from 55% (group 2) to 80% (group 5). Congestion of internal organs and congestion together with softening of the mucous membranes of the digestive tract with accompanying numerous hemorrhagic effusions were found in the dissected rats.

The distribution of radioselenium in the organs and tissues of the irradiated rats was, just as in the controls, a stepwise process, isotope concentrations varying from group to group and over different examination periods.

In the stomachs of rats treated with radioselenium 24 h (Tab. 1) after irradiation selenium concentration was within 2 days higher, then roughly similar, and after 24 days lower than in the controls. In rats of groups 2 and 3 small intestine selenium concentrations were lower than or similar to those in the controls. In groups 4 and 5 they were higher than those in the controls during the first 14 days and then declined to the control level.

Large intestine selenium concentrations after 1 day were in all rats higher than and after 14 days equal to those in the controls. Liver radioselenium concentrations were initially in all irradiated animals higher than in the controls and continued to be higher for 2 to 8 days. It was only in groups 4 and 5 that they afterwards dropped to the control level. Kidney concentrations were higher than

No.	2 d					4 d					8 d					14 d				
	C	52 mC	103 mC	155 mC	206 mC	C	52 mC	103 mC	155 mC	206 mC	C	52 mC	103 mC	155 mC	206 mC	C	52 mC	103 mC	155 mC	206 mC
1.	5·1	4·3	4·5	5·4	5·6	4·4	4·1	4·8	4·2	4·3	3·2	4·1	3·2	3·6	3·4	3·2	2·7	3·4	2·7	2·6
2.	10·0	9·6	9·6	13·0	14·6	4·9	6·4	6·5	6·5	7·3	2·9	4·1	2·9	3·0	3·2	2·6	2·8	2·8	2·2	1·9
3.	3·8	4·4	4·9	5·8	6·8	3·4	4·2	4·8	4·4	5·1	2·5	3·5	2·6	2·7	2·7	2·1	2·2	2·2	1·9	1·8
4.	13·0	16·8	14·4	20·1	20·7	11·2	15·7	17·5	14·0	15·9	10·3	11·4	10·3	10·0	10·4	7·3	9·1	9·5	7·6	7·0
5.	25·6	25·9	24·0	30·5	33·5	16·3	20·9	23·7	22·2	21·7	12·4	17·7	12·0	16·9	14·7	11·4	12·2	11·8	10·3	9·6
6.	3·5	3·8	3·0	4·0	4·2	3·5	3·9	3·8	3·2	3·2	2·7	3·9	2·5	2·7	2·8	2·8	3·1	3·0	2·4	2·2
7.	5·9	5·8	4·8	6·5	6·7	4·7	5·4	5·1	5·0	4·9	3·7	4·9	3·3	3·4	3·6	3·4	3·7	3·3	2·6	2·7
8.	1·4	1·1	1·1	1·4	1·2	1·1	1·2	1·2	1·1	1·1	1·0	1·2	0·9	1·0	0·8	0·9	1·0	1·1	0·9	0·7
9.	7·5	7·3	6·8	10·6	10·0	7·0	8·0	8·8	7·6	9·6	5·0	7·6	5·4	6·4	5·3	5·0	5·3	4·7	3·7	3·7
10.	7·5	7·8	6·3	7·8	9·0	5·9	6·0	6·3	5·3	5·0	4·0	4·9	3·3	3·8	3·6	3·8	4·2	3·6	2·7	2·3
11.	1·0	1·0	1·1	1·3	1·2	1·2	1·2	1·4	1·2	1·2	1·3	1·4	1·1	1·1	1·1	1·3	1·4	1·4	1·4	1·2
12.	8·9	8·6	7·9	9·2	8·9	9·9	11·0	12·2	11·1	10·7	11·1	16·2	11·6	13·1	11·9	13·3	15·9	17·1	14·2	13·4
13.	2·5	2·5	2·0	4·9	4·3	2·3	1·9	3·3	2·9	2·9	1·9	2·3	1·5	2·2	2·2	1·9	1·6	1·6	1·9	1·9
14.	2·6	2·7	2·2	2·9	2·8	2·5	2·4	2·8	2·5	2·2	2·0	2·2	1·7	1·8	1·6	1·4	1·7	1·6	1·5	1·1
15.	0·1	0·1	0·1	0·3	0·1	0·1	0·1	0·2	0·1	0·2	0·2	0·2	0·2	0·3	0·2	0·4	0·3	0·3	0·2	0·3

No.	2 d					4 d					8 d					14 d				
	C	52 mC	103 mC	155 mC	206 mC	C	52 mC	103 mC	155 mC	206 mC	C	52 mC	103 mC	155 mC	206 mC	C	52 mC	103 mC	155 mC	206 mC
1.	5·1	4·9	4·7	5·7	5·6	4·4	4·4	5·5	4·6	5·4	3·2	3·9	4·0	4·1	4·2	3·2	3·7	2·8	3·7	3·3
2.	10·0	11·9	12·2	11·2	12·8	4·9	6·0	5·9	4·7	6·5	2·9	4·9	3·7	3·8	4·4	2·6	3·2	2·3	2·6	2·6
3.	3·8	3·9	4·0	5·9	5·2	3·4	3·6	4·3	3·9	5·1	2·5	3·4	2·8	3·3	3·6	2·1	2·4	2·0	2·5	2·3
4.	13·0	13·9	16·6	16·4	16·3	11·2	14·4	15·1	13·3	16·6	10·3	13·9	13·4	13·5	12·3	7·3	8·4	8·8	9·2	8·9
5.	25·6	24·9	27·2	27·9	28·3	16·3	21·5	23·6	16·7	23·4	12·4	18·7	16·9	15·6	16·4	11·5	13·4	10·5	12·7	12·6
6.	3·5	3·8	3·9	4·5	4·2	3·5	3·8	3·7	3·3	3·7	2·7	3·6	2·5	3·7	3·4	2·8	3·3	2·6	4·1	3·2
7.	5·9	5·8	6·0	6·2	6·6	4·7	5·9	5·8	4·9	5·9	3·7	5·2	4·4	4·4	5·0	3·4	4·3	3·1	3·9	3·4
8.	1·4	1·3	1·2	1·6	1·3	1·2	1·0	1·2	1·0	1·1	1·0	1·2	1·0	1·1	1·1	0·9	1·1	0·8	1·1	1·0
9.	7·5	9·4	9·0	10·1	10·0	7·0	9·1	9·3	7·2	9·4	5·0	8·4	6·8	5·7	8·1	5·0	5·6	4·0	3·4	3·1
10.	7·5	8·5	8·2	9·0	8·3	5·9	6·8	6·8	5·0	6·2	4·0	5·7	4·0	4·7	4·3	3·8	4·7	3·6	3·9	3·7
11.	1·0	1·3	1·2	1·5	1·2	1·2	1·2	1·4	1·1	1·2	1·3	1·5	1·2	1·6	1·6	1·3	1·6	1·1	1·7	1·5
12.	8·9	9·2	9·2	13·0	9·9	9·9	12·4	13·8	10·7	12·7	11·1	19·3	15·2	16·9	16·5	13·3	18·9	14·6	21·5	17·3
13.	2·5	3·2	2·5	5·6	3·4	2·3	2·1	2·1	3·3	3·6	1·9	1·9	2·1	2·9	2·8	1·9	1·6	1·7	2·2	1·6
14.	2·6	2·8	2·6	3·4	2·8	2·5	2·3	2·6	2·4	2·6	2·0	2·9	1·8	2·2	1·9	1·4	1·7	1·2	1·8	1·4
15.	0·1	0·1	0·1	0·2	0·1	0·1	0·1	0·1	0·2	0·1	0·2	0·2	0·4	0·3	0·3	0·4	0·9	0·3	0·4	0·3

in the controls (group 1) and after 14 days they declined to become equal to those in the control group. The exception was group 5 where they dropped below the control level. Lung concentrations followed a similar pattern but after 14 days they were generally lower than in the controls. Heart and spleen selenium concentrations rose after 1 day and then declined to reach the control level after 8 days. Selenium concentrations in the blood were initially reduced. Selenium concentrations in the testis, prostate, skin, and hair varied slightly during the first 3 to 24 h. Within 14 days higher concentrations were found in the muscles but, as a rule, no changes were observed in the brain.

In rats treated with radioselenium 5 days after irradiation (Tab. 2) stomach selenium concentrations within 6 h after treatment were higher and then declined and stayed at the control level until the 14th day. Small intestine concentrations were increased for 8 days and on 14th day they dropped below the control level. Large intestine concentrations were increased after two days and stayed at an unchanged level until the 14th day. Liver radioselenium concentrations were in all rats higher than in the controls during 14 days and only in the animals of groups 4 and 5 they were lower during 6 to 24 h. Kidney radioselenium concentrations stayed higher than in the controls during 14 days in all animals. Heart and blood selenium concentrations were lower after 3 to 6 h, then after 1 to 2 days were higher than and afterwards equal to those in the controls. Lung and muscle selenium concentrations were during 14 days somewhat higher than in the controls. Prostate selenium concentrations were slightly increased compared to the controls but only after 1 to 2 days. It was only after 8 to 14 days that selenium concentration in the testes was found to be increased. Brain, skin, and hair selenium concentrations followed a pattern similar to those in the controls.

Discussion

The data presented in this paper suggest that the selenium concentrations in the organs of the control rats followed a pattern similar to those reported by *Magos* and *Webb* [25]. They also support the fact that selenium distribution in rats is similar to that in other laboratory animals [26, 27, 28]. The very high selenium concentrations in the small intestine are undoubtedly associated with selenium absorption. According to *Whanger* et al. [29], the duodenum is the main absorption site. That fact also affects selenium content in the liver since the latter organ performs detoxicating functions in the organism. It is in the liver that, according to *Caygill* et al. [30], selenite undergoes transformation into selenates and selenides. The highest concentration of selenium in the kidneys is related to the mode of its excretion from the organism. According to *Nakamura* et al. [31] in monogastric animals more than 60% of selenium is excreted with urine.

The clinical symptoms in the irradiated rats were typical of the X ray doses used. The changes in lymphocyte and leucocyte numbers were similar to those reported by *Dienstbier* et al. [32].

The changes in the distribution of selenium in the digestive tract of the irradiated rats were characterized primarily by an increase of the isotope concentration in the stomach, small intestine, and large intestine. Increased and somewhat variable selenium concentrations were found in the liver, kidneys, lung, spleen, heart, and skeletal muscles. There were no characteristic differences in those changes between the 1-day-after-irradiation treated rats and those treated 5 days after being exposed to X rays. Generally, those changes did not correlate with ionizing radiation dose. It should be stressed that in all those organs the selenium concentrations found after 14 days were similar to those in the controls. In the remaining organs (brain, testes, prostate, skin, hair) the distribution of selenium in the irradiated animals followed a pattern similar to that in the control group. The essence of those changes can be linked to the effect of ionizing radition on the regulatory function of the nervous system [33] and on functional morphology-related properties of organs differing in their sensitivity to ionizing radiation [34, 35, 36]. Postirradiation haemodynamic disorders can also be of significance in this respect [37, 38].

To sum up the present study it can be concluded that the effect of exposure doses (X rays) of 52, 103, 155, and 206 mC is confined merely to short-lived alterations of selenium distribution in some animal organs. Those alterations do not correlate with exposure doses. Selenium concentrations in the irradiated animals that occured 14 days after irradiation were found to be equal to or slightly lower that those in the control animals.

Received May 27, 1986

References

[1] *L. Dahle, E. G. Hill, R. T. Holman*, Arch. Biochem. **98** (1962) 253
[2] *S. W. Jordan, P. H. Dean, J. Ahlquist*, Lab. Invest. **27** (1972) 538
[3] *T. F. Slater*, Biochem. J. **222** (1984) 1
[4] *J. Luotola*, Med. Biol. **62** (1984) 115
[5] *Z. M. Bacq, A. Herve, J. Lecomte, P. Fischer, J. Blavier, G. Dechamps, H. Le Bihan, P. Rayet*, Arch. Intern. Physiol. **59** (1951) 442
[6] *H. Zalkin, A. L. Tappel, J. Jordan*, Arch. Biochem. Biophys. **91** (1960) 117
[7] *Z. M. Hollo, S. Zlatarov*, Naturwissenschaften **47** (1960) 328
[8] *H. S. Olcott, W. B. Brown, J. Vecu*, Nature **131** (1961) 1201
[9] *J. W. Hamilton, A. L. Tappel*, J. Nutr. **73** (1963) 493
[10] *A. T. Diplock*, Med. Biol. **62** (1984) 78
[11] *F. Shimazu, A. L. Tappel*, Radiat. Res. **23** (1964) 210
[12] *R. Badiello, E. Gattavecchia, M. Mattii, M. Tamba*, Biochem. Biophys. **29** (1979) 647
[13] *R. J. Shamberger*, Biochemistry of selenium, New York: John Wiley 1983
[14] *E. Cekan, B. Tribucoit, H. Vokal-Borek*, Acta Radiobiol. Oncology **24** (1985) 267
[15] *J. Rosenfield, D. A. Beath*, Selenium geobotany, biochemistry, toxicity and nutrition, New York: John Wiley 1969
[16] *S. Kossakowski, A. Kossakowska*, Medycyna Wet. **35** (1979) 108
[17] *W. K. Ermakov, H. W. Kowalskij*, Biologiczeskoje znaczenie selena, Moskwa 1974
[18] *R. J. Shamberger, E. Rukovena, A. K. Longfield, S. A. Tytko, S. Deodhar, C. E. Willis*, J. Nat. Cancer Inst. **50** (1973) 863
[19] *A. H. Cantor, M. L. Longevin, T. Nogucki, M. C. Scott*, J. Nutr. **105** (1975) 106
[20] *T. Westmark*, Acta Pharm. Toxicol. **41** (1977) 121
[21] *M. R. Gardiner*, Austr. Vet. J. **42** (1966) 442
[22] *G. Lunde, S. A. Oderyaard*, Nord. Vet. Med. **24** (1972) 484
[23] *E. L. Patterson, R. Mihtrey, E. L. R. Stockstad*, Proc. Soc. Exp. Biol. Med. **95** (1957) 617
[24] *K. Schwarc, C. M. Folte*, J. Am. Chem. Soc. **79** (1957) 3239
[25] *L. Magos, M. Webb*, Arch. Tox. **36** (1976) 63
[26] *S. Kossakowski*, Medycyna Wet. **36** (1980) 4
[27] *S. Kossakowski*, Medycyna Wet. **36** (1980) 500
[28] *F. Bernschneider, M. Hess, K. Neuffer, S. Willer*, Arch. Exp. Vet. Med. **30** (1976) 525
[29] *P. D. Whanger, N. O. Pederson, J. Hatfield, P. H. Weswing*, Proc. Soc. Exp. Biol. Med. **153** (1976) 295
[30] *P. J. Caygil, J. A. Lucy, A. T. Diplock*, Biochem. J. **125** (1972) 407
[31] *H. Nakamura, J. Sayato, M. Tonomura, G. Ose*, Chem. Abstr. **78** (1973) 155219 n
[32] *Z. Dienstbier, M. Arient, J. Pospizil, A. Smid*, in: Experimentalni postiradiacni syndrom, Praha 1966, p. 107
[33] *M. N. Livanov, D. A. Briyukov*, II United Nations Inter. Confer. on the Peaceful Uses of Atomic Energy, P/2315 – 1958
[34] *R. A. Conard*, Radiat. Res. **41** (1956) 167
[35] *Ch. E. Dunlop*, in: Pathology – *W. A. D. Anderson*, p. 219, 1961
[36] *H. Fritz-Niggli*, Radiobiologia, Warszawa 1965
[37] *S. Kossakowski, W. Łojek, J. Kuczyński*, Pol. Arch. Wet. **17** (1975) 4
[38] *L. G. Cockarham, T. F. Doyle, R. B. Trumbo, J. B. Nold*, Int. J. Radiat. Biol. **45** (1984) 65

Isotopenpraxis 22 (1986) 12, S. 451 bis 456

Examination of Arrangement of Alloy Structural Elements in Models of Metallographic Structures*)

M. Radwan, K. Gibas, K. Łybacka, E. Narbuntowicz

(Technical University of Poznań, Poland)[1]

Examining metallurgical structures by means of autoradiography one needs answering the question whether the geometrical shapes of component arrangements visible as dark spots are determining the type and parameters of probability distribution. Testing this question some autoradiographic models have been prepared representing different shapes of appearance of labelled components and 14 models presenting components of different size and density of appearance have been realized. The conclusions drawn from this studies will be used in further improvements of the description of alloy structure.

Bei der Untersuchung von metallurgischen Strukturen mit Hilfe der Autoradiographie muß die Frage beantwortet werden, ob die geometrischen Formen der Komponentenanordnungen, wie sie als dunkle Flecke sichtbar sind, auch wirklich der Art und den Parametern der Wahrscheinlichkeitsverteilungsfunktion entsprechen. Zur Testung dieser Frage wurden einige Autoradiogramm-Modelle durch Markierung der Komponenten hergestellt, die einer vorgegebenen Erscheinungsform entsprachen. Insgesamt wurden 14 Modelle mit unterschiedlicher Größe und Dichte der Erscheinungsbilder bearbeitet. Die erhaltenen Ergebnisse werden zur weiteren Verbesserung der Beschreibung von Legierungsstrukturen verwendet.

Keywords

alloys; autoradiography; distribution functions; materials testing; metallography; phantoms; radioisotopes; structural models

In our previous papers [1, 2] we have suggested a method of describing structure with probability distribution of concentrations of metal elements constituting an alloy. We began with the assumption that the term structure designates, most generally, a distribution of components (parts) together with a set of relations between them which are characteristic for a given object, arrangement, set of objects, system. Such a definition is consistent with many definitions given by various encyclopedia and dictionaries. Hence, constituent metal atoms concentrating in appropriate places of the alloy result in creating separate phases observable by means of a microscope. Optically distinguishable picture of the metallographic structure is usually correlated to alloy solidification system providing names to these phases and determining the mode of their occurrence, either qualitative or quantitative [3, 4, 5].

To verify the structure description method taking the probability of component concentration distributions in alloys into account, it is necessary to answer the following question: are the geometrical shapes of component arrangement visible on autoradiographs as the dark spots determining the type and parameters of probability distribution. To enable the determination like that, the autoradiograph models representing different shapes of appearance of labelled components have been prepared. 14 models presenting the components of different size and density of appearance have been realized. On these models the measurements of density according to a specially prepared net of random "addresses" have been carried out. The results of measurements formed the base to identify the probability distributions.

The following paper presents selected results illustrating distributions of black component probabilities on model autoradiographs.

Model "P 2", given in Fig. 1, has been photometered with an aperture 0·051 mm^2 according to a scheme of random "addresses" applied twice: straight on, and at 90° angle [6]. Data obtained from measurements on random places and fed into computer let us put forward a hypothesis that examined sample belonged to the population of beta distribution with the parameters $\gamma = 0{\cdot}896$; and $\eta = 2{\cdot}177$, and the function of probability density assumes the shape of upturned letter J (Fig. 2).

The second model from the group of autoradiographs with "spot" structure, "P 6" (Fig. 2), also comes from the beta distribution with similar parameters γ and η, and similar shape of function of probability density. Whereas from the set of measured data obtained from the "K 1" and "K 2a" models (this is shown in Fig. 3 and 4) of cellular structure it has been observed that the curve of probability density function assumes different same beta distribution family, but of course, with absolutely different parameters γ and η (Fig. 3 and Fig. 4).

Besides the cases presented here in detail also the measurements of another autoradiograph models have been done and comparison of the results is given in Tab. 1.

As results from these examples all empirical distributions obtained from the measurements are included in the family of beta distributions with different shape parameters (γ and η) and different curves of probability densities.

Since it has turned out that the beta distribution is of particular importance for autoradiographs being sets of

*) Presented at the XIVth International Symposium on "Autoradiography", 19–23 November 1984, Reinhardsbrunn, G.D.R.

[1]) Address: Institute of Machines Technology, Isotope Laboratory, PL-60699 Poznań, ul. Piotrowo 5

Model of autoradiograph P 2

Sample size: $N = 100$
Significance level: $\alpha = 0{\cdot}05$
Max. value of statistics K–S: $D = 0{\cdot}07800$
Critical value of statistics K–S: $\lambda = 0{\cdot}13600$

The examined sample at the significance level $\alpha = 0{\cdot}06$ is derived from the beta population with parameters $\gamma = 0{\cdot}896$, $\eta = 2{\cdot}177$

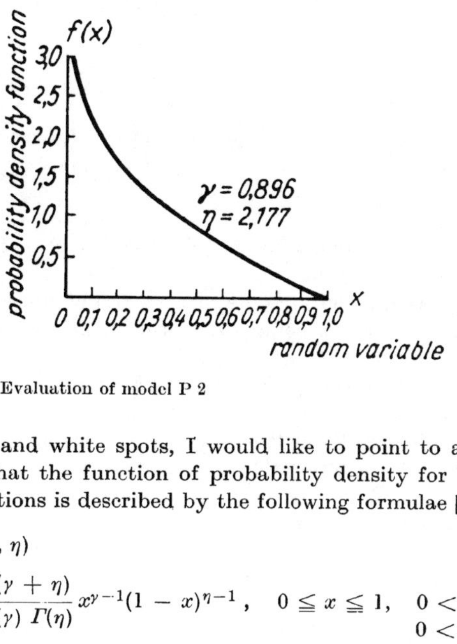

Fig. 1. Evaluation of model P 2

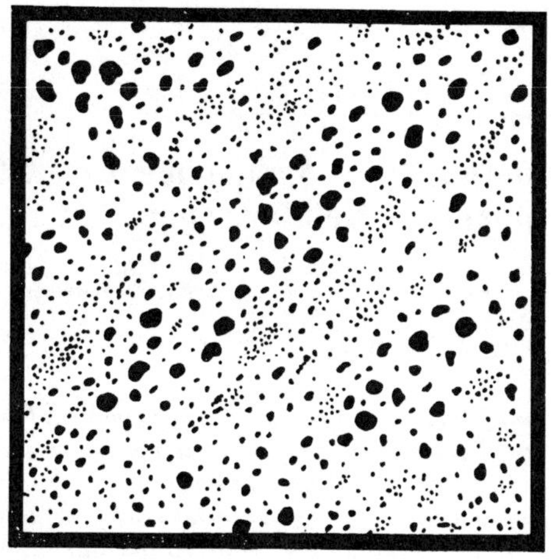

Model of autoradiograph P 6

Sample size: $N = 100$
Significance level: $\alpha = 0{\cdot}05$
Max. value of statistics K–S: $D = 0{\cdot}11900$
Critical value of statistics K–S: $\lambda = 0{\cdot}13600$

The examined sample at the significance level $\alpha = 0{\cdot}05$ is derived from the beta population with parameters $\gamma = 0{\cdot}646$, $\eta = 2{\cdot}674$

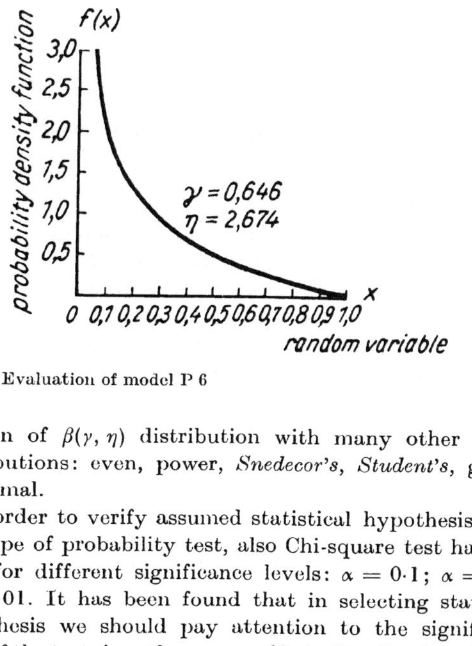

Fig. 2. Evaluation of model P 6

black and white spots, I would like to point to a known fact that the function of probability density for beta distributions is described by the following formulae [11]:

$$\psi(x;\gamma,\eta) = \begin{cases} \dfrac{\Gamma(\gamma+\eta)}{\Gamma(\gamma)\,\Gamma(\eta)} x^{\gamma-1}(1-x)^{\eta-1}, & 0 \leqq x \leqq 1, \quad 0 < \gamma \\ & \qquad\qquad\quad 0 < \eta \\ 0 \quad \text{elsewhere} \end{cases}$$

and the curves of this function take various shapes depending on the values of the two parameters γ and η:

1. When $\gamma > 1$ and $\eta > 1$ the distribution is single peaked with peak at
$$x = \frac{(\gamma - 1)}{(\gamma + \eta - 2)}.$$
2. When $\gamma < 1$ and $\eta < 1$ the distribution is U shaped.
3. When $\gamma < 1$ and $\eta \geqq 1$, the distribution is reverse J shaped, and when $\eta < 1$ and $\gamma \geqq 1$, it is J shaped.
4. When $\gamma = \eta$ the distribution is symmetrical.

Typical courses of the density function are given in the following figures (Fig. 5 to 8).

The importance of the beta distribution and particularly its possibilities of application, is greatly increased by connection of $\beta(\gamma, \eta)$ distribution with many other known distributions: even, power, *Snedecor's*, *Student's*, gamma or normal.

In order to verify assumed statistical hypothesis about the type of probability test, also Chi-square test has been used for different significance levels: $\alpha = 0{\cdot}1$; $\alpha = 0{\cdot}05$; $\alpha = 0{\cdot}01$. It has been found that in selecting statistical hypothesis we should pay attention to the significance level of the test since the curves of beta distributions suggest that the prognosed model can belong to normal, lognormal, exponential, gamma, etc. distributions.

Therefore tests should be applied assuming the highest significance level since the degree of uncertainty inherent to probabilistic model is reversely proportional to the number of data at our disposal. Thus the set of experimental data should be sufficiently large so that the differences between the parameters of each probability distribution were clear enough and the errors were the least.

Assuming that the structure is decided by local concentration of a given element, it is necessary to use correct technique of concentration measurement on the area which is the most appropriate for a given structure [2, 6]. Two techniques fulfil this requirement best: point X-ray analysis [7], and autoradiography [8, 9, 10]. Particularly recommended one is the radioactive structural analysis where analysed sample is irradiated with a homogeneous

Model of autoradiograph K 1

Sample size: $N = 100$
Significance level: $\alpha = 0.05$
Max. value of statistics K–S: $D = 0.08000$
Critical value of statistics K–S: $\lambda = 0.13600$

The examined sample at the significance level $\alpha = 0.05$ is derived from the beta population with parameters $\gamma = 1.244$, $\eta = 2.544$

Model of autoradiograph K 2a

Sample size: $N = 92$
Significance level: $\alpha = 0.05$
Max. value of statistics K–S: $D = 0.09000$
Critical value of statistics K–S: $\lambda = 0.13600$

The examined sample at the significance level $\alpha = 0.05$ is derived from the beta population with parameters $\gamma = 1.272$, $\eta = 3.962$

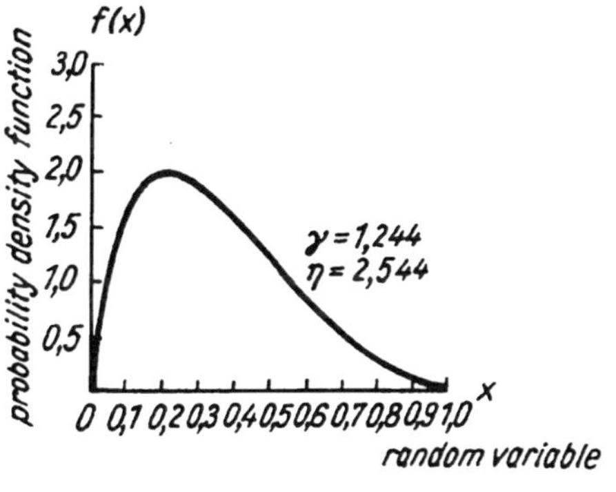

Fig. 3. Evaluation of model K 1

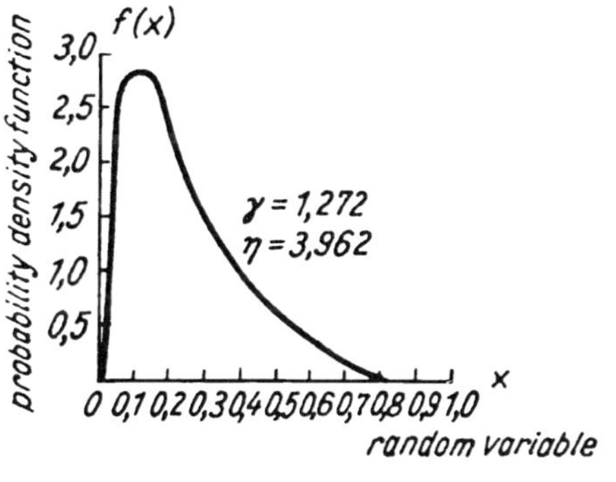

Fig. 4. Evaluation of model K 2a

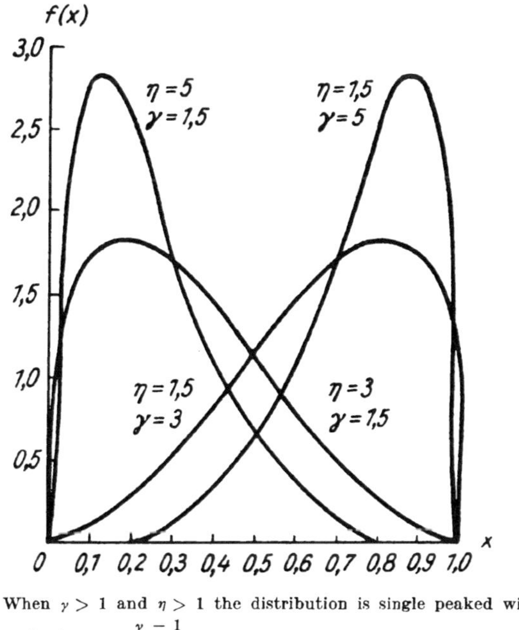

Fig. 5. When $\gamma > 1$ and $\eta > 1$ the distribution is single peaked with peak at $x = \frac{\gamma - 1}{\gamma + \eta - 2}$

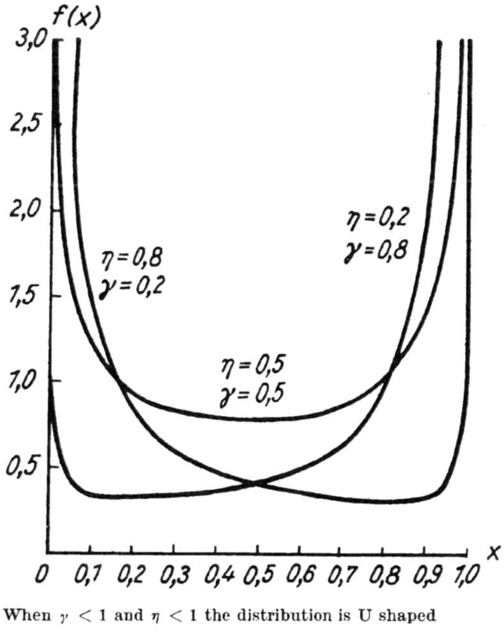

Fig. 6. When $\gamma < 1$ and $\eta < 1$ the distribution is U shaped

Tab. 1. The results of autoradiograph models evaluation

Model of autoradiograph	Sample size	Significance level	Mean value	Sampling variance	Coeff. of skewness	Coeff. of kurtosis (excess)	Parameters γ	Parameters η
K 1	100	0·05	140·64	5230·61	0·4953	2·1706	1·244	2·544
K 2	90	0·05	464·98	63885·96	0·9829	3·6260	1·199	3·893
K 2a	92	0·05	220·67	18908·50	1·3382	4·3466	1·272	3·962
K 2b	100	0 05	228·62	64534·05	1·7366	5·9599	0·457	2·347
P 2	100	0·05	61·31	1692·17	0·2279	2·1464	0·896	2·177
P 3	100	0·05	82·52	7752·13	1·1667	3·7302	0·451	1·986
P 4	99	0·05	123·71	14359·08	1·4021	4·5822	0·536	2·612
P 6	100	0·05	70·45	3884·01	1·7491	5·5834	0·646	2·674

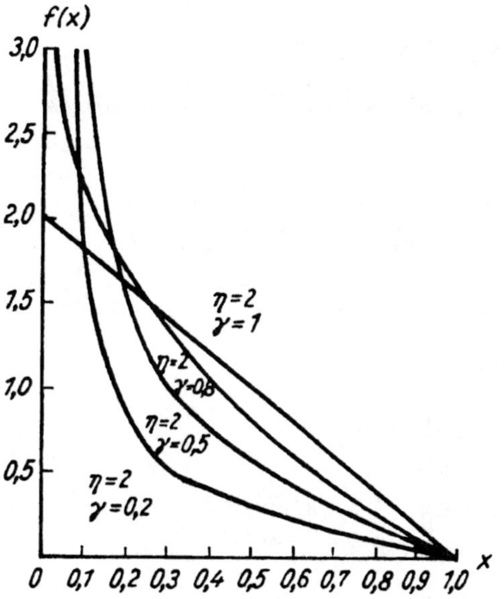

Fig. 7. When $\gamma < 1$ and $\gamma \geqq 1$ the distribution is reverse J shaped, and when $\eta < 1$ and $\gamma \geqq 1$, it is J shaped

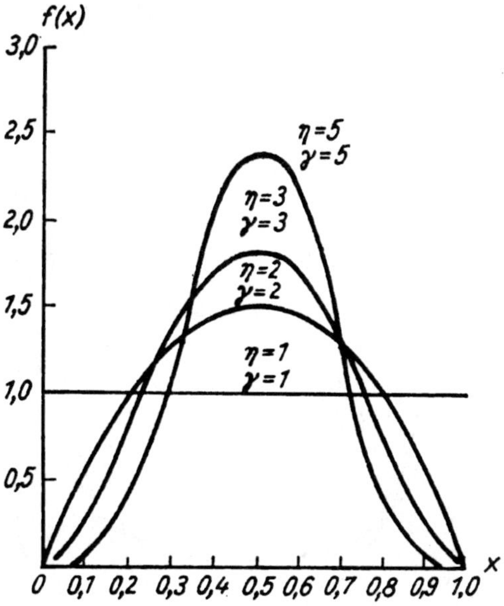

Fig. 8. When $\gamma = \eta$ the distribution is symmetrical

stream of charged particles or neutrons [12]. Due to very pronounced differences occurring between cross-sections of isotopes of different elements for radiation escapement of neutrons, and between their half-life periods, other elements have only very slight negative influence on results of the analysis. Thus in alloys of elements showing these differences we can obtain the radioisotope which represents examined element. This method was applied in presented research work whose aim was to verify the method of probabilistic description of the alloy structure of one of the most interesting arrangement of alloy solidification — a system with peritectic. The peritectic appears in many systems such as Cu—Sn, Cu—Zn and Cu—Al. However, the intermetallic phases are observed in these systems.

As we have stated, the intermetallic phases may make impossible to indicate correlation between solidification system and the probability distribution of alloy component concentration.

The Pt-Ag alloys are the only ones solidifying according to the system with peritectic in which no other phases appear (Fig. 9 to 11). The Pt-Ag alloys were examined (simple peritectic system) for which the difference between ordinal numbers Z for Pt and Ag, and large cross-section of Ag activation $n - \gamma$, as well as long half-life period of 110m Ag (270 days), facilitate obtaining autoradiograms of Ag distribution in the alloy structure. The radioactivation structural analysis has been applied the activation of specimens from which the autoradiographs have been taken was carried out in the neutron stream of reactor

$$\Phi = 10^{11}\ \text{n cm}^{-2}\ \text{s}^{-1}$$

during 1 hour.

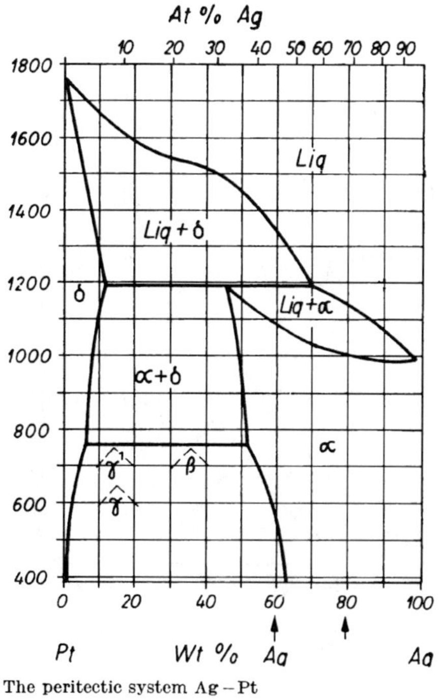

Fig. 9. The peritectic system Ag—Pt

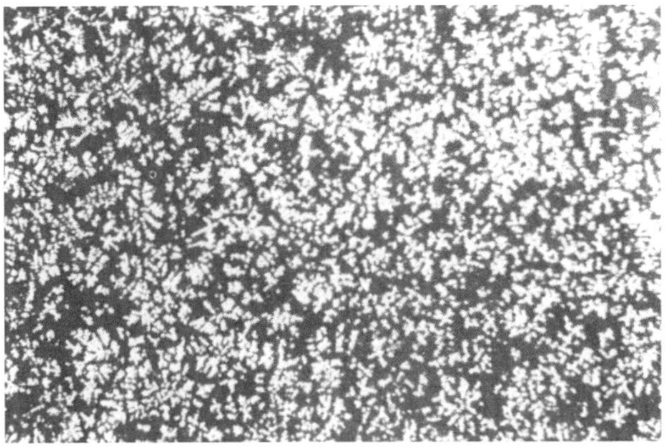

Fig. 10. Structure of Pt-wt% 60 Ag alloy (100 ×)

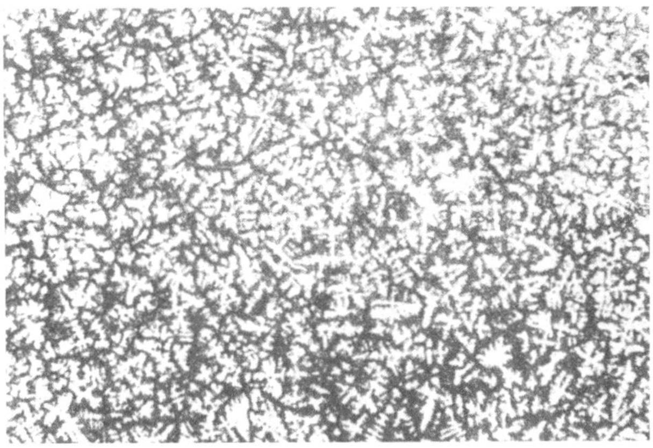

Fig. 11. Structure of Pt-wt% 80 Ag alloy (100 ×)

Data are listed in Tab. 2.

Two alloys have been examined using the autoradiographic method and electron probe microanalysis:

a) Pt-wt 80% Ag,
b) Pt-wt 60% Ag.

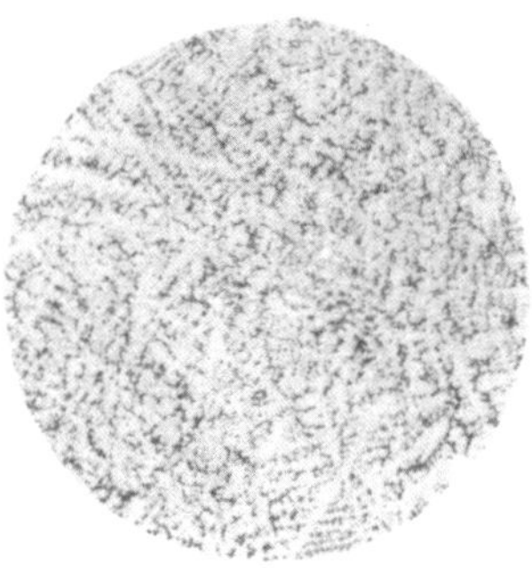

Fig. 12. Autoradiograph of Pt-wt% 60 Ag alloy

Fig. 13. Autoradiograph of Pt-wt% 80 Ag alloy

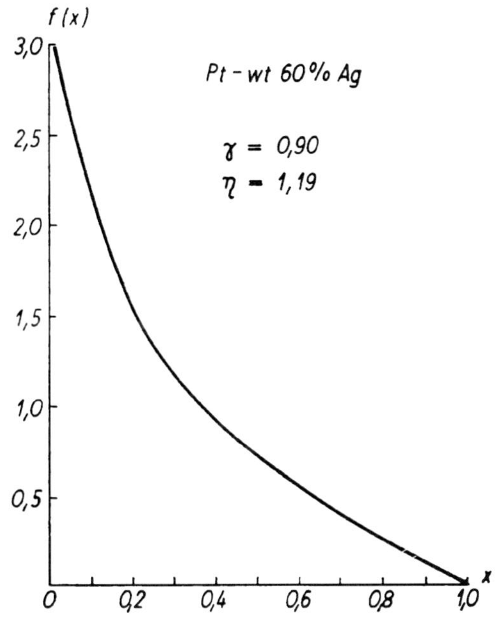

Fig. 14. The curve of probability density function for Pt-wt% 60 Ag alloy

The autoradiographs have been photometered (Fig. 12 and 13). The results of measurements of autoradiograph density as well as those for silver concentration in platinum have been used to define the type and to determine the values of probability distribution parameters for the component concentration in alloy.

The results show that the different structures have a different distribution of the same element (Fig. 14 and 15) and this was a main argument for taking into account the probabilistic method of evaluation of alloy autoradiographs.

Received October 16, 1985

Tab. 2. Parameters of activation (n, γ) of isotope elements in Ag-Pt alloys

Element		^{47}Ag	^{78}Pt			^{77}Ir	^{44}Ru
Mass number of isotope		110 m	193 m	191	195 m	192	103
Activity [kBq] in 1 g alloy	Ag-wt 40% Pt	67·5287	0·1021	0·0104	0·1465	0·3223	0·0118
	Ag-wt 20% Pt	90·0769	0·0511	0·0067	0·0777	0·3223	0·0118

(n, γ) reaction, $\Phi = 10^{11}\ n \cdot cm^{-2} \cdot s^{-1}$ $\tau = 1$ h, $t = 720$ h

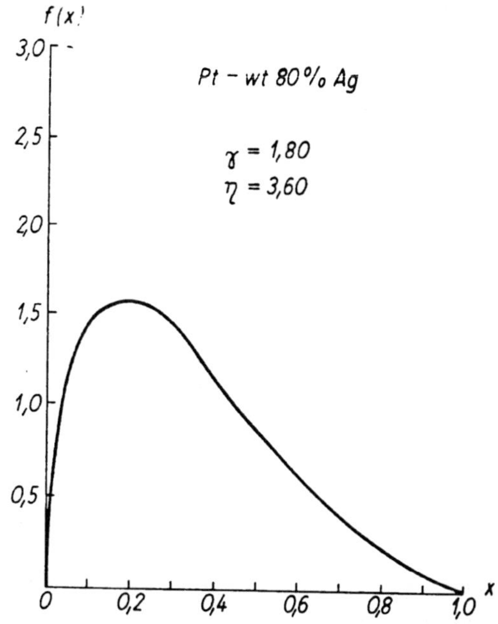

Fig. 15. The curve of probability density function for Pt-wt% 80 Ag alloy

References

[1] *M. Radwan, K. Gibas, E. Skowrońska*, Isotopenpraxis 18 (1982) 6

[2] *M. Radwan, K. Gibas* – Proceedings of the XIIIth International Symposium on Autoradiography, Tábor, Czechoslovakia, 3 – 5 May 1983, pp. 63 – 69, 107 – 113

[3] *S. A. Saltykov*, "Stereometrische Metallographie", Leipzig: VEB Deutscher Verlag für Grundstoffindustrie. 1974

[4] *J. Ryś*, "Metalografia ilościowa", Kraków: Skrypty Uczelniane AGH 1965

[5] *J. Sokołowski, K. Hübner, M. Nosila, B. Pluta*, "Metalografia ilościowa", Gliwice Skrypty Uczelniane Politechniki Slaskiej 1979

[6] *M. Radwan*, Isotopenpraxis 21 (1985) 141

[7] *J. Lubkiewicz*, "Badania w zakresie zastosowań ilościowej mikroanalizy rentgenowskiej w metaloznawstwie". Warszawa 1976

[8] *S. Z. Bokszlein, S. T. Kiszkin, L. M. Moroz*, "Badanie struktur metali za pomoca izotopów promieniotwórczych", Warszawa: WNT. 1962

[9] *M. Radwan, J. Sarna*, "Izotopy promieniotwórcze w badaniach metaloznawczych", Warszawa: WNT. 1970

[10] *M. Radwan, A. L. Jankowski, J. Kominek, K. Szalzchneider*, "Autoradiografia metaloznawcza". Warszawa 1968

[11] *J. Hahn, G. Shapiro*, "Statistical Models in Engineering", New York: John Wiley, 1967

[12] *M. Radwan*, Isotopentechnik 2 (1962) 9

[13] *G. J. Daries*, "Solidification and Casting". London 1973

Isotopenpraxis 22 (1986) 12, S. 456

Buchbesprechung

Ecological Studies – Analysis and Synthesis

W. D. Billings, F. Golley, O. L. Lange, J. S. Olsen, H. Remmert (editors), Vol. 55, A Eutrophic Lake, Lake Mendota, Wisconsin, Th. D. Brock. Berlin – Heidelberg – New York – Tokyo: Springer-Verlag 1985. 308 S., 82 Abb., 47 Tab.; geb. DM 148.–

Was ist das für ein See, über den ein ganzes Buch geschrieben und von dem behauptet wird, er wäre einer der am meisten untersuchten Seen überhaupt? Auch in einem guten Atlas findet man die vier kleinen Seen, an welchen die Hauptstadt des Bundesstaates, Madison, liegt, nur als kleine Punkte – wenn überhaupt –, und einer davon ist der Mendota-See. Der Name stammt aus dem Sioux-Indianerdialekt und bedeutet so viel wie „der Mund des Flusses". Er hat eine mittlere Tiefe von 12,4 m, eine Fläche von 39,4 km^2 und ein Volumen von 48,10^7 m^3, ist also wirklich ein kleiner See, ungefähr so groß wie der Kummerower See oder das Steinhuder Meer oder halb so groß wie der Chiemsee. Was ihn aber auszeichnet, ist die Tatsache, daß sich seit Ende des vergangenen Jahrhunderts Limnologen und Süßwasserbiologen bemühen, den See so gründlich wie nur möglich zu untersuchen, angefangen durch *E. A. Birge* (1851 – 1950) und fortgeführt von *Ch. Juday* (1871 – 1944). Im Jahre 1889 wurde zum ersten Male die Wasserblüte des Sees beschrieben, was ihn als Lehrbeispiel für einen eutrophischen See auszeichnet und Grund genug ist, ihn in einer Monographie exzeptionell ausführlich darzustellen.

Neben einer physisch-limnologischen und chemischen Charakterisierung des Sees, die sich vor allem durch Langzeitreihen auszeichnet, sind längere Abschnitte den biologischen Eigenschaften und Besonderheiten (Phytoplankton, Zooplankton, höhere Stufen der Nahrungskette) gewidmet, die es erlauben, Modellrechnungen der Stoffkreisläufe des Sees auszuführen. Als Hauptursache der Eutrophierung wird die in der Umgebung des Sees betriebene Landwirtschaft, weniger die Folge der Urbanisierung in den letzten Jahrzehnten, angegeben.

Von den Isotopentechniken wurden die ^{14}C-Methode zur Bestimmung der Photosyntheseraten oder der Methanoxidationsraten und die ^{35}S-Methode zur Bestimmung der Sulfatreduktionsrate im anaeroben Bereich des Sees eingesetzt. Man vermißt allerdings die Messungen der Isotopenvariationen von C, N und S bzw. die ^{210}Pb-Methode zur genaueren Bestimmung von Sedimentationsraten. Hervorzuheben sind die tabellarischen Zusammenstellungen der Meßdaten, die eine vertiefte Analyse auch in anderen Zusammenhängen gestatten.

Das Problem der Eutrophierung von Seen kann kaum in einer Seminarstunde ausreichend diskutiert oder gar verstanden werden; die vorgelegte Monographie erlaubt Lernenden und Lehrenden ein umfassendes Studium des Erscheinungsbildes der Nährstoffanreicherung am Beispiel eines kleinen Sees und macht deutlich, wie weit wir noch von flächendeckenden und trendanalytischen Aussagen entfernt sind.

H. Hübner

40 Jahre
wissenschaftliche Literatur
im Dienste des Friedens und des
gesellschaftlichen Fortschritts
Philosophie/Ökonomie
Geschichte
Alte Geschichte/Archäologie
Literaturwissenschaft
Sprachwissenschaft
Physik
Mathematik
Geowissenschaften
Biowissenschaften/Medizin
Chemie
Zeitschriften
AKADEMIE-VERLAG
THEORIA CUM PRAXI
BERLIN
Deutsche Demokratische Republik
1946-1986
DDR-1086 Berlin, Leipziger Straße 3-4 · Postfach 1233

ISOTOPENPRAXIS

Band 22 1986 Heft 12

Contents / Inhalt

►Language of Publication ►Publikationssprache

Isotopenpraxis is indexed in Current Contents/PC & ES, Science Citation Index, and INIS Atomindex

FSC
www.fsc.org
MIX
Papier | Fördert gute Waldnutzung
FSC® C083411

Zeitfracht Medien GmbH
Ferdinand-Jühlke-Straße 7
99095 Erfurt, Deutschland
produktsicherheit@kolibri360.de